An OPUS book

The Philosophies of Science

D0060785

Rom HARRÉ

The Philosophies of Science

Second edition

Oxford   New York

OXFORD UNIVERSITY PRESS

1985

Oxford University Press, Walton Street, Oxford OX2 6DP

Oxford New York Toronto
Delhi Bombay Calcutta Madras Karachi
Kuala Lumpur Singapore Hong Kong Tokyo
Nairobi Dar es Salaam Cape Town
Melbourne Auckland

and associated companies in
Beirut Berlin Ibadan Nicosia

Oxford is a trade mark of Oxford University Press

First edition published 1972 as an Oxford University Press
paperback and simultaneously in a hardback edition
Paperback edition reprinted five times
Second edition published 1985 as an Oxford University Press
paperback

British Library Cataloguing in Publication Data

Harré, Rom
The philosophies of science. — 2nd ed —
(OPUS)
1. Science — Philosophy
I. Title   II. Series
SO1 Q175
ISBN 0–19–289201–0

Library of Congress Cataloging in Publication Data

Harré, Rom.
The Philosophies of science.
(An OPUS book)
Bibliography: p.   Includes index.
1. Science—Philosophy. 2. Science—History.
3. Philosophy. 4. Metaphysics.
I. Title. II. Series: OPUS.
Q175.H3264 1984   501   85-13846
ISBN 0–19–289201–0 (pbk.)

Printed in Great Britain by
The Guernsey Press Co. Ltd.
Guernsey, Channel Islands

Preface to Second Edition

IN THIS BOOK I have tried to introduce the philosophy of science in such a way as to bring out the way problems which appear specific to science are actually species of wider philosophical issues.

There have been many attempts to make sense of the methods by which we acquire our knowledge of nature, and to set that knowledge within a more comprehensive metaphysics. While I have aimed at making manifest the variety of ways in which a rational basis for science has been sought, I have also tried to show how, in the end, two opposed positions seem to crystallize out of the apparent variety. There is the positivist position, which tends to treat theories as if they are mere logical structures, efficacious only for making predictions. With this goes a tendency to restrict scientific knowledge to generalizations about the passing show of sense-experience. Over against this stands the realist point of view which emphasizes the work of the human imagination in leading to conceptions of the real ties behind sense-experience, and which admits the content of theories to the status of empirical knowledge.

I have tried to present the case for positivism as fairly as possible, though I believe the case against that point of view, on intellectual, historical, and moral grounds, to be overwhelming. In recent years a new threat to scientific realism has emerged. The identification of social influences on the formation and assessment of theories by sociologists of science has led some philosophers to propose a radical relativism. In this new edition a short introduction to this fascinating development has been added.

Preface to First Edition

IN THIS BOOK I have tried to introduce the philosophy of science within wider philosophical contexts, in the attempt to show how the problems which appear to be specific to science are actually species of wider philosophical issues.

There have been many attempts to make sense of the methods by which we acquire knowledge of nature, and to set that knowledge within a more comprehensive metaphysics. While I have aimed at making manifest the variety of ways in which a rational basis for science has been sought, I have also tried to show how in the end two opposed positions seem to coalesce out of the apparent variety. There is the positivist position, which tends to treat theories as if they are theorems in geometry, and to restrict empirical knowledge to the passing show of sense-experience. Over against this is the realist point of view which emphasizes the work of the human imagination in leading to conceptions of the realities behind sense-experience, and which admits the content of theories to the status of empirical knowledge.

I have tried to present the case for positivism as fairly as possible, though I believe the case against that point of view, on intellectual, historical, and moral grounds, to be overwhelming. The bibliography at the end of this book contains suggestions for further reading along both realist and positivist lines, which I hope will allow the reader to form his own opinions on these difficult matters.

Contents

List of Figures

1

The Philosophy of Science

MOST PEOPLE SUPPOSE that philosophers think about very general and very deep questions, at the heart of which is the problem of the relation of Man to the Universe. Philosophers are popularly thought to offer ideas about the general purposes of living, and even the more particular aims one should set oneself in one's ordinary life. Philosophy of science in this sense would be a discussion of the place of the scientific enterprise in the whole pattern of life. It would probably be concerned with providing an ultimate justification for doing science, that is, with whether science was worth doing at all. It might be argued, for instance, that the accumulation of scientific knowledge is destructive of the conditions for living the best possible human life. It might be thought that the effort expended in the pursuit of scientific knowledge might be better employed in the cultivation of artistic sensibility, in the refinement of manners, and in the embellishment of the environment. I am not going to pursue that kind of discussion, though I am very far from thinking that discussions of such general questions have no value.

In this book I shall be discussing a great many detailed questions which arise in the actual practice of science itself. I shall be trying to explain how our knowledge of the world and the things in it is advanced. I shall be trying to make clear what principles are assumed in the use of time-honoured methods of acquiring knowledge. We shall find that certain principles are operative in scientific work. It is the aim of this book to make these principles manifest.

Philosophy has four main branches: *logic*, the theory of reasoning; *epistemology*, the theory of knowledge; *metaphysics*, the theory of concepts and their relations; and *ethics*, the theory of evaluation, particularly moral evaluation. I shall not be concerned here with ethics. We begin, in this chapter, with a preliminary account of the first three branches of study, and briefly discuss the relations between them. By the study of examples of the investigation of logical, epistemological, and metaphysical questions our understanding of these branches of philosophy will deepen.

Logic

Logic is the study of the canons or principles of correct reasoning. To discover logical principles from the study of examples we need to be able to recognize when a piece of reasoning is correct. If we knew, only by reference to principles of logic, which arguments were valid and which invalid, then we should have no need to study examples of reasoning to try to discover principles, because we should already know them. In fact we are able to tell whether some piece of reasoning is correct or incorrect without knowing or deliberatively applying any principles of logic, that is, without explicit reference to any canons of correct reasoning. The study of logic will enable us to say why some piece of reasoning is correct or incorrect. However, once the principles of logic have been extracted from examples it is inevitable that they should be used as canons, that is to express the standards to which reasoning should conform.

We must beware of supposing that the principles of correct reasoning, say in mathematics, hold good for other subject matters, say chemistry. This would be like supposing that all languages really have the same grammar. Believing that, one might be tempted to argue, for example, that English nouns really do have case endings, but that these are implicit, or understood. But it would be just as reasonable an alternative to ask oneself whether perhaps the application of the grammatical category 'case' to English nouns is misplaced. In this book I shall make no assumptions whatever about the transferability of principles of logic from one field to another. In particular I shall not assume that the logical principles appropriate to mathematics must be appropriate to the methods of reasoning in every natural science.

The written expression of scientific knowledge ideally takes the form of a reasoned and systematic exposition. Conclusions will be backed up by reasons. Hypotheses will be considered with respect to the balance of favourable and contrary evidence. Certain logical relationships hold between conclusions and the reasons for the conclusions. Other logical relationships hold between hypotheses and the reasons which call for their rejection, or modification. Relationships such as these are the stuff of logic. They must conform to the canons or principles of correct reasoning. Some considerations do support a conclusion; others do not. Judgements on such matters as these express the logical commentary upon the discourse. Logical principles must be produced to back them up. And they, in their turn, can come under critical scrutiny.

Let us now look at an example of scientific reasoning. In § 944 of his *Experimental Researches*, Michael Faraday wrote, apropos of a number of experiments which show 'the power of heat to produce a current by exalting the chemical affinity': 'I cannot but view in these results of the action of heat, the strongest proofs of the dependence of the electric current in voltaic circuits on the chemical action of the substances constituting these circuits: the results perfectly accord with the known influences of heat on chemical nature. On the other hand, I cannot see how the theory of contact can take cognizance of them, except by adding new assumptions to those composing it.'

The steps in his reasoning can be set out as follows: If current is produced by a chemical action, then increase in that action should increase the current; we know that increase in heat increases chemical action, so application of heat should increase the current produced. It does increase the current produced, so chemical action must be responsible for the current.

Are you convinced? Is Faraday's reasoning correct? The argument can be further analysed:

Heat increase causes increased chemical action
Heat increase causes increased electrical activity

therefore

Chemical action causes electrical activity.

Or still more schematically:

> If p, then q
> If p, then r
> ---
> If q, then r.

Is this argument form valid? Is it convincing? Is it correct or incorrect, as it stands? Would setting this argument in a wider context of experiment and theory change the way you feel about it?

This example is an analysis of the structure of an evidential relation, since in the argument Faraday is advancing reasons for accepting a hypothesis. This example has a very simple structure. We shall be concerned later with the more elaborate structures of theories and explanations. We shall try to find the principles according to which good theories are constructed.

In some sciences, like astronomy and meteorology, the making of predictions is a very important part of the work of scientists. We shall follow the ways in which predictions are made, and look into the means by which our confidence in them is justified. Logical principles are involved in the making of predictions. But certain conditions have to be satisfied in order to apply them. Consider what is involved in predicting an eclipse of the moon. First the general laws of lunar and solar motion must be known. An astronomer must also know where the sun and moon have appeared in the sky previously. In order to be justified in applying the laws of solar and lunar motion he must believe that the past behaviour of these heavenly bodies is a good guide to their future behaviour, that is, he must believe that they will continue to seem to move as they have always seemed to move. This belief involves the assumption that the sun and moon are stable material things. Thus a number of assumptions of different degrees of generality are involved in prediction. The acceptance of all these assumptions is a necessary condition for the application of the logical schema:

> *From* Previous positions of the sun and moon
> *And* The laws of lunar and solar motions
> ---
> *Infer* Times and places of lunar eclipses.

Epistemology

Epistemology is the theory of knowledge. In epistemological investigations we reflect on the standards to which genuine knowledge should conform. We try to characterize the kind of knowledge which a given method of study might yield about a certain sort of subject matter, and how far that kind of knowledge conforms to what are taken to be standards of genuine or true knowledge. From these considerations we may be able to form some idea of what sorts of facts we could never get to know. It is the job of the epistemologist to show how knowledge can be distinguished from true belief, and certainty from probability.

This study is an important part of the philosophy of science. Philosophers of science are interested in determining how far confidence in particular methods of discovery should extend. They are also concerned with more general epistemological questions, such as whether knowledge of the existence of things and materials is more certain than knowledge of the effects that things and materials have upon our senses. Philosophers and scientists would like to know whether there is any part of scientific knowledge that is certain and not liable to revision under any conceivable circumstances. There are many other important epistemological questions which will be discussed in later chapters of this book. For example: 'How do new discoveries affect the status of what we already think we know?' 'Is the information acquired by learning a theory different in kind from that acquired by making an observation?' 'Can observations be made without a scientist having some theory in mind?' 'Is all knowledge, in the last analysis, theoretical knowledge?' The discussion of each of these questions will raise other questions, some of which are epistemological, but others will lead us into logic and metaphysics.

As a brief, introductory example of the discussion of an epistemological question, consider how we might answer the question: 'What do we *really* know?' This is not so easy to answer, because it is possible to cast doubt on the most certain-seeming pieces of information, and on apparent matters of fact. Let us take two different kinds of case.

First, suppose you ask yourself: 'Am I absolutely *sure* that I am, at this moment, looking at the *page* of a *book*?' Doubt can be cast by asking how one knows that the rest of the book exists, when one is looking only at one page. It is not impossible that one should turn over the page and find a hollow space instead of the other pages one normally expects. This is a hypothesis that is open to empirical test, and that doubt can be settled. Other doubts of the same kind can be raised. Perhaps the print is being cast upon a blank page by a skilfully concealed projector. Perhaps the print is being cast upon a blank page by the 'reader's' powers of eidetic imagery. All these doubts can eventually be settled.

But there is another kind of doubt. It is conceivable that you are now dreaming that you are holding a book. In what way could you *prove* that this is not so? I am inclined to think that there is a sense of proof in which neither you nor anyone else could prove to themselves that they were really reading a book and not dreaming. One could not prove it as one could prove a theorem in geometry. However, one is entitled to say one *knows* one is reading a book because making sure one has a book before one is not at all the same sort of thing as proving a theorem. One can prove that it is a book by the usual ways in which one distinguishes real books from imaginary books, and from books in dreams. Should we conclude from this that the doubt I have been raising is absurd? It does, in a way, show this, because proving that one is really reading a real book does not require or admit of the rigour of geometrical proof.

The point of pressing these doubts is to show the extent of the assumptions that are made in our treatment of the world. These are empirical assumptions. We assume that books are physically the same all the way through and we assume that the print is on the page. There are also metaphysical assumptions. To assume that not all our experiences are dreams is a metaphysical assumption. It does not admit of empirical test. It is really an assumption about the conceptual system we are to adopt. To treat the problem about the reality of dreams on a par with the problem about the reality of the inside of an unopened book is to slip from a reasonable scepticism about matters of fact to an almost wholly pointless terminological issue. Dreaming is a state identified by contrast with being awake. If we are persuaded to call all our experience 'dreams', then we will have to introduce a new pair of terms to mark the old distinction between dreaming and being awake, because we shall still have to distinguish between *waking* dreams and *dream* dreams. 'Dreaming' would now be used for all our experiences, and would no longer mean 'dreaming'. It would mean something like 'experiencing'. That is, it would cover both our present state of dreaming *and* our present state of being awake. Of course there is sometimes point in a philosopher proposing such a terminological revision. The new use of 'dreaming' is neither the old use, nor is it quite synonymous with what was previously meant by 'experiencing'. It leaves one with the feeling that experience is generated from within ourselves, and is not just the effect upon us of contact with other things and creatures. Some shift in vision occurs with this persuasive new way of using the word 'dream' and perhaps does change our ideas of experience.[1] It is clear, I hope, that it would be a most serious mistake to rate our belief in our being awake, when we are awake, as an assumption on a par with the assumption of the uniformity of material in a book, or the assumption of the stability of the seasonal weather pattern, and so on.

Let us ask again, 'What do we *really* know?', in different circumstances. If all scientific knowledge derives from observations of what happens in our immediate vicinity, how can we be properly said to know anything about distant regions of space and time? Again, if observations are confined to what a scientist can sense—that is, see, feel, touch, taste, and so on—what are we to make of the information that is conveyed by a chemical equation? Does it really describe the distribution and redistribution of atoms in molecules, happenings which cannot be observed? How can we be properly said to know anything about the behaviour of things too small to be observed? If we accepted these doubts a chemical equation would simply be a summary description of the changes in colour, taste, texture, and so on of the materials which had taken part in the chemical reactions that had been seen on earth, and the changes in the distributions of the weights of the substances that had occurred in earthly laboratories and factories. The view that this was what chemical equations expressed was held by Sir Benjamin Brodie in the 1860s. We shall examine it in more detail later. In investigating the force of these doubts epistemological questions have to be answered. Do theories offer a special kind of knowledge different from the knowledge we get

[1] For a general discussion of this 'shift of vision', see F. Waismann, *How I see Philosophy* (London: Macmillan, 1968), ch. 1.

from observation and experiment? What sort of knowledge do we get from instruments?

In the philosophical study of instruments, we investigate the differences between the kind of knowledge obtained from using instruments which improve and extend our senses, like the microscope, and the kind obtained from using instruments which detect phenomena which we could not observe since we lack the necessary senses. A paper covered with iron filings, or a little compass, can be used to detect a magnetic field. In describing what we learn from the use of detecting instruments, is it right to say that we have discovered facts about a magnetic field *as well as* facts about the way iron filings and compasses behave in proximity to an electrified coil of wire? Did the discovery of ultra-violet and infrared radiation just extend our knowledge of colours?

These questions lead in the end to queries about the ultimate subject matter of scientific laws. Various answers to this question will be considered in the course of the discussion. Some people have thought that, strictly speaking, the content of scientific laws should be considered to be confined to the statistics of sets of numbers derived from the readings of instruments. Others have thought that the laws of nature are about the behaviour of real things and materials that make up the world as we know it. Yet others have thought that they described nothing but the ordered sequences of sensations that we experience. Instead of treating the laws of geometrical optics as being about the passage of light rays through different systems of media, they would treat them as describing the sequences of luminiferous sensations in our visual fields.

Metaphysics

These days metaphysical studies are more modest than they have been in the past. No one of any discretion writes about the Universe, Man, and God. In modern metaphysics the most general concepts used in science and in ordinary life are investigated. For example a modern metaphysician might study the space-concepts and time-concepts used in ordinary life and compare them with those used in Special Relativity. He might examine various concepts of cause, or the concepts of possibility and necessity. Modern metaphysics is aimed at achieving clarity of thought by a careful study of concepts. In part this is done by a study of various aspects of language use. A modern metaphysician will try to discover how various concepts are related. He might investigate how our thing-concepts are related to our space-concepts. He might consider whether our concept of temporal direction is contingently or necessarily connected with our concept of causality. In recent years some of these conceptual problems and studies have moved into the foreground of science. Discussions of problems about the concept of causality and the concept of thinghood have appeared in physics, particularly in commentary upon attempts to interpret quantum mechanics in a useful and consistent way. Conceptual problems about space and time have cropped up in the relativity theories. Problems as to what are limits of the concept of an individual have cropped up in biology, particularly in discussions of the attempts to specify the unit of evolution, that is to say *what* evolves. The recent dissatisfactions in psychology partly revolve around feelings of uncertainty as to the metaphysics of concepts such as 'person' and 'act'.

The relation between logic and epistemology

In the philosophy of science the three branches of philosophical study cannot be studied profitably in isolation from each other. Solutions offered to problems in one field inevitably affect the kind of solutions that are then possible in others.

Suppose that in considering the logical relations between the evidence for a law and the law, it is decided that it is not proper to speak of *inferring* the law from the evidence. This solution might be adopted because it is thought that that way of speaking suggests the existence of a logical relation where none exists. The statements describing the evidence will be *particular* statements, of the form: 'In this experiment, when light passed from one medium to another the sine of the angle of incidence bore the same ratio to the sine of the angle of refraction as it did in the other experiments I have done.' The law will be expressed in a quite *general* statement: 'In *all* cases of the passage of light from one medium to another the sine of the angle of incidence will bear a constant ratio to the sine of the angle of refraction.' Snell's Law is quite general. But deductive logic does not sanction an inference from particular to general. So to speak of *inferring* the law from the evidence is at least misleading, since it tempts us to think that the law is a conclusion drawn from particular premisses according to some principle of logic. If the principles of deductive logic do not govern the extraction of laws from observations, then we cannot say that the truth of laws follows, with logical rigour, from the truth of the statements describing the evidence for them. We might draw an epistemological conclusion from this, namely that while we can be said to know the truth of statements describing particular experiments, we cannot be said to know the truth of the Laws of Nature, which we suppose to be based upon that evidence in some way. We might go so far as to argue that because of this it is misleading to speak of the truth of laws at all. Perhaps they should be spoken of as more or less satisfactory conjectures, thus leaving the way open for new and conflicting evidence to appear. It might be argued that while we can be said to know those particular facts about the world which make up the evidence for laws, we can only properly be said to have a *belief* in a law. We should not say that we *know* laws, because that implies that the law is true. One cannot know what is not true.

Logical considerations affect epistemology in the problem of the status of predictions. Can we ever be certain of the truth of a prediction? Astronomical prediction comes as near to certainty as we can attain, so it will serve as an example. The position of a planet can be determined with great accuracy. Statements describing the positions that it has occupied serve as premisses from which its future positions can be inferred with logical rigour. Nautical almanacs and tables of ephemerides are made up of the results of just such inferences. We might be tempted to think that the predictions were just as certain as the data upon which they were based. It is tempting to think that *accuracy* of observation automatically yields improved certainty of prediction. Accurate observations are essential to accurate prediction, but certainty is another matter. The certainty of a conclusion is the certainty of the weakest link in the chain of deduction, that is, of the least certain premiss. If I am quite certain that John has red hair, and I am fairly confident that all red-haired people are Celtic, then I can be only fairly confident that John is a Celt.

To use the astronomical data to make an astronomical prediction by deduction, an astronomical theory is required. That theory, consisting in part of laws, will be general. According to the arguments above, it cannot be known to be true. At best it can be said to be the most satisfactory theory yet devised. The possibility that it might be abandoned altogether means that it is less certain than the astronomical data. Looking at the matter in this light we would be inclined to rate the theory as the weakest link in the chain that leads from data to prediction. But theory is sometimes more certain than data. If we compare our confidence in the correctness of a theory with our confidence in the data of observation and experiment, I believe that there are very many cases where we should be more inclined to favour the theory. For example, the theory of evolution is generally held to be correct though the data upon which it is based are pretty unreliable and incomplete. We shall come later upon a further complicating factor in this matter. Our views as to the reliability of the data, and even to some extent how the data are interpreted, depend upon the theory that is held by the investigator. There is no such thing as pure data. Whichever way we look at it, however, predictions based upon data and arrived at by the use of a theory are no more certain than the certainty of the most dubious of the elements entering into a prediction. If a prediction is not fulfilled, and we are confident of the correctness of the data, and of their interpretation, then the theory will have to be revised, and perhaps in extreme cases even overthrown.

This kind of uncertainty about the future becomes less as we discover more facts, and refine and elaborate our theories. We can be a great deal more certain now about the future positions of the planets, about the effects of temperature on chemical reactions, about the progress of a patient with a certain disease than we ever could before. And our certainty in such matters must surely increase. But there is a philosophical trap here for the unwary. It is very easy to be pushed from the reasonable concession that we can never be absolutely certain now about the future course of events, to quite unreasonable doubt about the possibility of any knowledge of what is to come. By asking ourselves whether we *really* know what is likely to be the outcome of some process we can easily slip into thinking that we do not really *know* what is likely to happen. If we do not *know* that carbon dioxide will turn lime water milky, we must be ready for the possibility that the next time we pass carbon dioxide through lime water anything may happen. It may turn green. In the course of these few sentences the

reasonable doubts we may have about the precise outcome of experiments and exact fulfilment of predictions are being turned
into scepticism, by an unnecessarily stringent condition being put
upon what is to count as knowledge. If all we are entitled to say
we know must be absolutely certain, then of course we have no
knowledge of the future, and no knowledge of anything else. But
the distinction between having knowledge of the future and merely
guessing is one of the contrasts by which the concept of knowledge
gets its sense. Scientific studies do not produce information which
is absolutely certain. But science is not guess-work. Indeed, in the
stringently constrained sense of knowledge in which we have no
knowledge of the future, only the truths of mathematics are
knowledge.

The relation between logic and metaphysics

The above were examples of the influence of logic on epistemology.
There are also cases in which ideas derived from epistemological
reflections have influenced logic. We shall run across some of these
cases later on. Now I want to look briefly at some examples of the
interactions between logic and metaphysics. The style in which
logical analyses of propositions are made can influence one's views
as to the ultimate metaphysical categories. One way in which
logicians analyse general statements is by the use of predicates and
a logical device called a quantifier. An expression such as 'is
gaseous' is called a *predicate*. Common nouns can be replaced by
predicates without any apparent loss of meaning. Instead of 'This
horse is brown', I could say 'This animal is equine and is brown.'
An expression such as 'All' or 'Some' is called a quantifier by
logicians. It is usual to employ variables x, y, etc., with quantifiers
in the following way: instead of 'All horses are brown', we write
'All which are equine are brown', eliminating nouns in favour of
predicates. Then instead of the vague 'which', we use a variable, x,
and rewrite the sentence yet again as

For all x, if x is equine, then x is brown.

This sentence means the same as 'All horses are brown.'

Using this method of analysis on the law that all gases have the
same coefficient of expansion, one arrives at the sentence

For all x and for all y, if x is gaseous and y is gaseous, then the rate at which x expands = the rate at which y expands.

Instead of being about materials, that is, gases, the new form of the law seems to be about the properties of being gaseous and of being capable of expansion. An expression couched in terms of nouns seems to be about substances and things, while an expression in terms of predicates seems to be about qualities and processes.[1]

Adherence to predicate logic may lead one into thinking that a proposition is expressed better in a sentence using only predicates than in a sentence using nouns and adjectives. It is easy then to go on to think that the predicate mode of expression is somehow a more accurate reflection of the way things are, truer to nature as it were, than the use of nouns. This feeling could be expressed in a metaphysical theory that 'property' and 'quality' are more fundamental categories than 'substance' and 'thing', and that things should be treated as collections or collocations of properties and qualities. On this view, an apple is not *something* which is sweet, red, and round, but is the nexus of the qualities, sweetness, roundness, and redness. Are things nothing but collocations of qualities? This is a difficult question and its resolution is a problem in metaphysics. We cannot decide the question scientifically. Whatever experiments we could do the question could still be raised. It is really a conceptual question. Are thing-concepts analysable without remainder into conjunctions of quality-concepts?

Some metaphysical questions are close to scientific questions and problems, and are connected with them. Science does not consist only in doing experiments. Scientists are also involved in developing an adequate and a self-consistent system of concepts with which to understand the world as revealed in the results of experiments. In most sciences thing-concepts abound. What role do they play? Are they eliminable? Are the conditions for the application of thing-concepts violated by the concepts needed for subatomic physics? What assumptions do we build into our conceptual system just by using nouns like 'neutron' and 'proton'? By considering the nature of the thing-concept some part of an answer can be found, since the use of nouns and the employment of thing-

[1] For a general discussion of the issue between nouns, predicates, and quantifiers, see W. V. Quine, *From a Logical Point of View* (Cambridge, Mass.: Harvard University Press, 1953), ch. 1.

concepts are two aspects of the same metaphysical commitment. To these questions too we shall return in our detailed discussion of the philosophical problems of science. We shall see not only that logic influences metaphysics but that metaphysics has profound effects upon logic also.

The relation between epistemology and metaphysics

I have already mentioned the epistemological theory according to which all we can claim really to know is that we are presently having particular sensations. According to the theory, knowledge is confined to such facts as that I am now experiencing feelings of pressure in my fingers, and we can only infer that these derive from my holding a pen. That there is now a white patch in my visual field and that there is now a buzzing in my ears are other such facts. We have seen that this view is based upon the extravagant exploitation of perfectly reasonable reservations about the possibility of absolutely certain knowledge of the world. It seemed, however, that, whatever one might doubt, one could not doubt the truth of present ascriptions of current sensations and feelings to oneself. Instead of talking of the *paper* before me, which involves dubious assumptions about the existence of the reverse side of the sheet, and of the permanence of the paper in time, I can talk instead about a white patch in my visual field, and this way of talking carries no assumptions about the existence of reverse sides of sheets, or even of the existence of the sheet when I am not aware of the paper, for instance when I am not looking at it or feeling it. White patches in visual fields do not have a back and do not exist at times other than those at which they are experienced. Instead of talking of the pen in my hand I can talk of feelings of pressure in my fingers. So, it has been argued, for scientific purposes, where absolute certainty seems to be an ideal, the subject matter of study should be confined to those items about which one can have absolute certainty, that is to one's own sensations. On this view a thing should be treated as nothing but a co-presence of looks, feels, tastes of a certain kind. If I am not touching a thing, then on this view to say that there is a thing which I am seeing or hearing is to imply that under certain circumstances I would experience a feeling of resistance in my finger-tips. The thing-concept seems to be replaceable for scientific purposes by conjunctions of actual and potential sensation-concepts.

A whole metaphysical system can be erected on this basis. It has been called 'Phenomenalism'. All the key concepts of science can be reinterpreted within this theory. The fact that changes in one thing can bring about or cause changes in another, is treated as reducible to the fact that a sensation of one particular kind is usually followed by a sensation of another particular kind in people's experience. Spatial relations between things are analysed as temporal relations between feelings. For example, the distance of one thing from another could be treated as the number of the kinaesthetic feelings associated with stepping-out that I have between one touch sensation and another. Successive sensations in time become the ultimate realities. This is a *metaphysical* doctrine, and it is often supposed to be supported by the epistemological theory that all we can know for certain is our present having of sensations. Later we shall take a more detailed look at philosophies of science which are connected with this epistemological viewpoint.

The consequences for epistemology of the adoption of a metaphysical theory can also be profound. A striking and dramatic example of this is the Special Theory of Relativity. Like all important theories in physics it is a blend of metaphysical and empirical elements. The empirical element is the alleged fact of the constancy of the velocity of light in all inertial frames. This is the hypothesis that the velocity of light will be the same whatever be the relative velocities between the bodies upon which it is measured. For example, that velocity will be the same for light coming from bodies towards which the earth is moving as for light coming from bodies from which the earth is moving away. The metaphysical element in the theory amounts to a denial of the intelligibility of *empirical* concepts of absolute position and time. Any system of objects has equal title to the claim to be 'absolutely at rest'. Whichever is chosen as reference frame, the motions of all other systems can be determined with respect to it. But since *any* other system of objects can be chosen as the frame of reference, the motions determined with respect to the first choice have no special title to be described as absolute. So there is no way of telling which motions are absolute. No empirical concept of absolute motion could have application. The theory of relativity assumes that there is absolute simultaneity between events, but, according to the theory, we can never determine which events are simultaneous with which, without reference to some arbitrarily chosen frame of reference. So though the concept of absolute simultaneity is intelligible it cannot be given empirical application. This is the originality of the theory. The concept of absolute simultaneity is excluded from science in favour of an empirical concept which makes possible consistent judgements of simultaneity once a choice has been made of which objects to suppose to be at rest. The metaphysical theory about the kinds of concepts of motion and simultaneity to admit to science, together with the empirical assumption of the universal constancy of the velocity of light, leads almost directly to various epistemological theses, such as the thesis that whatever may be the actual relation of events in time we can never *know* which are absolutely simultaneous at different places.

I hope it has become clear that philosophical problems are assumed in any scientific practice. We have to choose some concepts with which to think about the world, and this amounts to devising or learning a language, and accepting a system of picturing and conceiving the structures in the world. Any set of concepts we choose, no matter how much they may lack systematic connection, involves metaphysical, epistemological, and logical assumptions. If we choose to employ thing-concepts we are already embroiled in a metaphysics that assumes the continuity of individuals in time (that is, that individuals endure through certain minor changes), because that assumption is an essential feature of the use of thing-concepts (that is, that an entity endures in time is one of the necessary conditions in which it would properly be called a thing). One aim of philosophy of science is to make these kinds of assumptions explicit, by exploring the concepts which are being used, so as to see exactly what is involved in using them, and to discover if they fall into some kind of system. One must try to get clear just what concepts are being employed in any given intellectual enterprise.

The value of this study for science itself derives from the additional powers that accrue to a scientist when he knows explicitly what assumptions are involved in the language and models he uses. If the assumptions are known they can be changed systematically, explicitly, and controllably. No amount of experimental work alone can determine what concepts are best to use, because to make an experiment already requires some formulation of a problem, and this requires the use of some concepts. This is not to deny that some are more suitable than others. For instance the laws of gaseous chemistry, discovered in the early years of the nineteenth century, were strikingly characterized by the fact that the proportionality relationships discovered were expressible in integral proportions; one volume of one gas combined with two volumes of another to give two volumes of a third, and so on. It is not at all surprising that an atomistic conception of matter began to be associated with these discoveries, though there were many chemists who disliked such a conception, and many who were able to feel satisfied in rejecting atomism altogether as a conceptual system suitable to chemistry.[1] On the other hand, there can be little doubt that the philosophical theory of primary and secondary qualities, according to which certain qualities of bodies were held to be really known to us in our sensory experience, while other qualities of bodies manifested themselves in experience quite differently from their nature in things themselves, led to a conceptual system, Corpuscularianism, which profoundly influenced the pursuit of science, and influences it still.[2] But whatever may be the major direction of influence at any time, the explicit identification of the structure and components of one's conceptual system release one from bondage to it.

The foregoing is a sketch, and only a sketch, of some of the fields of study of philosophy of science. The next step in our investigation will be to set out in a broad general way what science is, what scientists are trying to do, and how they go about their task. By this means we shall acquire a vocabulary for talking about the scientific process and its products. We shall then return to the fields of logic, epistemology, and metaphysics, to examine some of the great classical theories concerned with scientific knowledge, and what we think of as its possibilities and its limits in actual historical contexts with respect to actual pieces of scientific research.

[1] D. M. Knight, *Atoms and Elements* (London: Hutchinson, 1967).
[2] For an account of Corpuscularian philosophy and natural science, see R. Harré, *Matter and Method* (London: Macmillan, 1964), pt. II.

The world as it is, and the world as it is perceived

Suppose we were to ask a tolerably well-informed layman what he thought scientists did. Probably he would say that they studied various things to try to find out how they behave, how they work, and what they are. Some are studying viruses, others study rocks, some study the behaviour of jelly-fish, others study the anatomy of such creatures, yet others study the stars, and chemists and engineers study materials to try to find out their composition and how they will behave in different situations. Some of these materials are ordinary stuffs with which we are pretty familiar; others are more rare. Studying such things and stuffs involves analysing them, testing them, stimulating them, and then writing up any discoveries with a theory to explain the origin, or the behaviour, or the composition of the various things studied. The totality of things and materials is the world. Science has practical results and applications because once we know enough about the things of the world and the materials of which they are made we can do all sorts of things and bring about all sorts of effects which we may envisage as desirable. For example, it is now a commonplace that only when we really understand the nature of a disease are we really in a position to control it.

For philosophers the first and most elementary distinction of all is between the world as it is, and the world as it manifests itself in its effects upon sensitive objects—that is, as it is perceived by us, and detected by instruments. There are metaphysical theories which would argue that there is no distinction between these alleged two aspects, that the world is just as it manifests itself in its effects upon sensitive objects. But I am going to brush all theories aside for the moment and take a simple-minded view. After all, people are perfectly capable of distinguishing between the world as it is and the world as it is manifested to them. This capacity is shown in our capability of recognizing and making allowance for illusions. We know that distant people appear smaller than people close up. But we have grown so accustomed to allowing for this that our judgements of the relative sizes of people and things allow for their relative distances from us without explicit recognition of the facts of perspective. Kepler tells, with some glee, the story of the sleepy coach passenger who mistook a spider hanging from the top of the coach window for a peculiar ox in a nearby field. We can perfectly well make many such distinctions between appearance and reality.

Observation and detection

To return now to consider the effects of the world upon sensitive objects. There are undoubtedly two kinds of such effects.

First, there are the effects upon people in the course of which some philosophers and most laymen are inclined to say that the people are experiencing sensations and having feelings; I am going to use the general term 'sensation' for describing this kind of effect upon a person. Sometimes the sensitive object is not just a person, but a person plus an instrument, of the special kind I shall call 'sense-extending', like a telescope, a miscroscope, a probe, a stethoscope, and similar equipment, which allows someone to feel something which is not at the end of his finger-tips, or to see something which he could not otherwise see, that is, could not see without his spectacles or telescope.

The other sort of effect which things have is upon those instruments which are not sense-extending. An example of this kind of instrument is the electroscope. The instrument is set up so as to be protected from draughts. Two gold leaves hang down from a central rod which is connected to a copper disc. If an electrified rod is brought up to the disc the gold leaves diverge. This instrument is clearly sensitive to electrification. But it is not a sense-extending instrument. There is a curious prickling sensation that we experience when near an electrified thing, but we are not then perceiving an electric charge as a space-occupying entity. Our bodies are acting as electroscopes. An electroscope does not extend our senses in the way a microscope does. It is a device for *detecting* something imperceptible. The detection of a magnetic field by the use of a compass needle is even further removed from perception, since we have no awareness at all of being in a magnetic field.

The nature of experience and assumptions

People very rarely describe their experiences in terms of sensations. It is most unusual for anyone to speak of coloured patches passing across their visual field. If one does speak in this way it is to a doctor, and we expect him to try to do something about it.

Our experience comprises such perceptions as of things in action, horses running for example, and materials undergoing processes, such as water coming to the boil. We are aware of things in relation to other things, for instance of one horse pulling up on others against which it is racing.

A horse-race seems a commonplace enough happening, yet metaphysically it is a rather complicated affair. There has to be a relatively permanent material frame of reference constituted by the course. There are moving material objects, the horses, which must continue to exist from the start of the race to the finish. Just think of the assumptions involved in believing that the horse which has gone out of sight behind a large person who is blocking the view of the course is the same horse that is still leading after the horses emerge into view again. What would you say to a bookmaker who persisted in doubting that the horse which finished first was the one upon which one had bet at the start of the race because it had not been under continuous observation all the time? What would you say to a physicist who insisted that electrons existed only at such times as they were interacting with instruments?

To use the concept of a thing it is necessary to assume the existence of one's 'things' even when they are not being observed or detected. The justification of such an assumption leads one into a whole constellation of metaphysical problems. The world as perceived by us is made up of a variety of entities including things and materials with processes occurring in them. Our ordinary language allows us to say that we can feel, hear, taste, touch, and see things and materials undergoing processes and in changing relations to each other.

Percepts and sensations

Sensations have been supposed to be the ultimate units into which our experience can be analysed. When I perceive a thing, say a cup, the experience is supposed to be analysable into a group of elementary sensations such as coloured patches in my visual field and feelings of pressure in my finger-tips. Sensations are the contribution of the world to my experience. A cup, as I perceive it, is seen and felt by me as a solid object having a kind of independence, capable of occupying different places and of enduring in time. These features of my perception are not sensations, and, some philosophers believe, are not effects of the world upon me. Some philosophers believe that the organization of sensations into percepts, of coloured patches and felt pressures into cups, is something imposed by the person who has the experience. Sensations are supposed to be the ultimate units of which percepts are composed as well as the ultimate units into which all our experience can be analysed. A similar treatment is often proposed for processes, for which the elementary units of composition and analysis are events such as the having of a sensation by a sensitive organism. We shall examine a philosophy of science based upon the idea that science is really concerned with the study of sensations, and not with the things which cause us to experience those sensations. This is the philosophy of science associated with the metaphysical theory we have identified as phenomenalism.

There is some contribution from the knower to the known. Some of that contribution occurs at the level of experience. I shall argue that the most important contribution comes *from* the scientist *to* his scientific knowledge when he is expressing this knowledge in descriptions of observations and theories. It comes through the language he uses for these purposes. In describing what we see we use the language of things which endure in time. In explaining the happenings of which we are aware we use the language of causes, that is we describe things and materials acting upon each other. This is true, too, for the language we use for describing and explaining the behaviour of instruments. The behaviour of an instrument such as an electroscope, is described in the language of things and causes. We do not describe an electroscope as merely flipping up its gold leaf, we speak of it as detecting an electric field. We pass in thought through the raw effect to some substantial reality by speaking in this way. An electric field is something that is distributed in space and endures in time and has causal powers. It will cause things to happen other than the divergence of the leaves of an electroscope. Described in these terms, an electric field begins to gain substance in our thinking. We begin to treat the electroscope as a device which detects something real, not just as something which reacts to the presence of an electrified rod.

Some pretty deep philosophical problems can be raised about the assumptions involved in my distinction between the world as it is, and the effects it has upon us and our instruments. How can we be sure how far the world, as we perceive it, is the world as it is? How can we be sure that our detecting instruments are detecting substances or states of things of which we are not otherwise aware? Perhaps the real world is quite different from the world as we perceive it. Indeed, it might be quite different from what we surmise it to be by our understanding of our instruments. Perhaps we never get into contact with the world as it really is. Then, of course, our *problem* withers away, because the concept of the 'world as it really is' has no application, and we can get on with studying the world as it seems to be.

There is another problem of greater interest. Suppose that, impressed by the idea that what we perceive is in part dependent upon how we understand the world, and that there is more to seeing than having light impinge upon the eye, we try to achieve a true objectivity by aiming only at recording the sensations we are having when we are perceiving things, materials, and processes. Does true objectivity lie this way? It seems that true objectivity could not lie in the direction of what we perceive because we have conceded that perception is a species of understanding, rather than a simple effect of the world upon our sense organs. We shall return to the problem of objectivity and sensations from time to time. It might also be asked what is the justification for interpreting the behaviour of instruments on the model of seeing, hearing, and touching, that is on the perceptual model. Doubts might be cast upon this conception if the ideas of imperceptible things and processes which are supposed to explain the behaviour of instruments are thought to be mere fictional conceptions, mere theory, by the assumption of which some kind of order can be brought into the results of making experiments with instruments. I shall not resolve this doubt now, but return to consider it from different points of view as our elucidation of the philosophical problems of science proceeds.

The world as manifested to us is some kind of product of the operation of our understanding and the effect of the real world upon us. If we cannot provide a foolproof recipe for separating the one element from the other, how far can we rely upon the world as manifested being in any way like the world as it is? This is the general form of the problems considered above. However, the natural sciences with their techniques of model building are not dependent upon the assumption that the world as manifested is identical with or even very like the world as it is. Indeed, in the sciences we construct a picture of the world as it is that is consciously different in several ways from the world as it is seen, touched, heard, and tasted. Absolute objectivity of *observation* is not a possible ideal of science.

Thinking and experimenting

This leads naturally to a consideration of the intellectual as opposed to the observational part of science. Though we know that these are not truly separate parts of the work of scientists, we can turn our attention to the part of science that is done by thinking, imagining, writing, speaking, drawing diagrams, and so on, without supposing that making observations and using instruments is an activity devoid of intellectual content. Scientists not only offer us descriptions and classifications of things and materials and their actions and interactions, but also give us explanations. They can often explain why there are the things that there are. They can often explain why things and materials behave the way they do. Such explanations are usually given by the formulation of a theory.

A theory is expressed in sentences, diagrams, and models, that is in verbal and pictorial structures. Logic is the general theory of verbal structures, and the theory of iconic models is the general theory of pictorial structures. Logic and the theory of iconic models are both involved, as we shall see later, in the organization of sentences in all the intellectual activities of scientists. Various false pictures of these activities will entice us as we proceed. One is the idea of science as the accumulation of separate truths by the addition of one fact to another fact, each independently verified by experiment. This is perhaps the most commonly held view, and will be the first to which we shall turn our critical attention.

Scientific work is as much a work of the imagination as it is work at the laboratory bench. It is by the aid of disciplined and rationally controlled imagination that hypotheses as to the nature of things are invented. We must usually first imagine the mechanisms which produce their behaviour and which alone can suggest fruitful lines of further study. Science is not natural history, it is not the accumulation of facts. It is the building of a picture of the world. It is an intellectual enterprise aimed at understanding the world. What makes it different from other such enterprises, say that of the makers of works of art, is that it is done under the discipline of the experimental method. And this discipline is severe. A theory whose consequences are not borne out by experiment and observation must be modified, or some defect in the experiment demonstrated. Nor is science just the devising of mathematical laws and systems of laws which adequately cover the numerical results of quantitative experiments. The pursuit of this ideal alone leads us straight into triviality. A theory must serve as the basis for explanation, it is not just a codifying device. In order to fulfil this demand, a theory must describe the means by which the phenomena it explains come about. A theory must refer to the mechanisms of nature, not just to the quantitative results obtained by studying those mechanisms in action. Such results belong to the disciplinary part of science, rarely to its creative side.

Description and explanation, observation and theory, seem to be neatly distinguished aspects of the same distinction. What is observed can be described, if the resources of language are equal to it, and typically one makes use of theory in giving scientific explanations. It is hardly a scientific explanation of phenomena merely to describe some other phenomena with which they are associated, unless one has some conception of how this association comes about. Then that conception is what is really doing the explaining and is the heart of the theory. For instance, it is not a scientific explanation of the Aurora to instance the increased activity of sun-spots which regularly antedate the appearance of the glow in the sky. A scientific explanation will tell why and how the sun-spots are associated with the Aurora, and this involves discussions of the nature of sun-spots and of the paths of electrons which leave the sun. These discussions are relevant only because we have some idea about the nature of the Aurora, and know a good deal about the discharge of electricity in tenuous gases. In short, to explain the Aurora we describe the mechanism which produces the phenomenon, and so come to see *why* sun-spots are associated with the Aurora. In so doing we mention lots of things other than sun-spots and the Aurora. The differences between describing phenomena and explaining them, and between explaining them and describing the conditions under which they occur, will make themselves more and more clearly apparent as detailed examples unfold in later chapters.

But insisting upon the differences between observation and theory, and between description and explanation, must not be allowed to obscure the extent to which they are interrelated. Descriptions of observations cannot be entirely independent of theory either in form or in content. There are no modes of description which remain invariant under all changes of theory. The way in which observations are described changes where theory changes. The accepted way of explaining phenomena enters into the very meaning of the terms used to describe them. It seems to be generally agreed among philosophers, now, that the ideal of a descriptive vocabulary which is applicable to observations, but which is entirely innocent of theoretical influences, is unrealizable. In understanding a description we need to be aware of the current explanations of the phenomena described. Both scientific and metaphysical theories enter into descriptions and into our understanding of them in several ways.

The involvement of theories in description

Metaphysical theories: the use of nouns and adjectives in describing a phenomenon tends to bring along with it the metaphysical theory of substance and qualities, and thus entails the assumptions of that theory. We have seen how the elimination of nouns in favour of predicates would involve a profound shift in metaphysical assumptions, and could lead to the idea that the category of substance was less fundamental than the category of quality. Of course, one does not have to make this metaphysical move; one might regard the elimination of nouns in favour of adjectives as just a matter of linguistic technique.

The involvement of scientific theories in the meaning of descriptive terms is very obvious in the case of metaphors, a common and important part of descriptive vocabularies. The use of the term 'current' in describing electrical phenomena is not innocent of overtones of the fluid theory of electricity. But other, more neutral-seeming terms bear in their meaning some elements of a theory too. To speak of 'reagents', 'catalysts', and 'compounds' in chemistry is to use a vocabulary whose terms derive at least part of their meaning from the atomic theory of chemical change. To understand the *description* of a substance as alpha-hexachlorobenzene involves an understanding of the structural theory of molecular organization. I think it not too much to say that there is no term used in describing observations upon the nature and behaviour of things and materials, whose meaning is able to be grasped without some knowledge of the relevant theory or theories. In short, in learning to apply descriptive terms we are also learning theories. If you have learnt to identify pyridine only by its smell, then you do not really know what the word 'pyridine' means.

Examination of the idea that theories are not descriptive

Theories clearly differ from descriptions of the observations they are meant to explain. But does this difference imply that the terms used in a theory are not descriptive, and cannot significantly be used to refer to actually existing things, states, materials, and processes other than those they are explaining? Consider a term like 'atom', which is always cropping up in theories in physics and chemistry. Thousands of kinds of observations are explained by reference to atoms. Without this notion, physics and chemistry would be totally different in content. And yet no one has ever observed an atom, with or without sense-extending instruments. Single atoms have neither been seen, heard, touched, felt, nor tasted. Are there then really any such things? Certainly the epistemological point that we cannot know about them, as individuals, by being acquainted with them directly must be conceded. Does it follow from that that we cannot know about them at all? If this were so we should have to say that the word 'atom' did not mean what we had thought it meant. We had thought that 'atom' meant the smallest chemically independent piece of material. We have been using the word 'atom' so that if we saw and touched a piece of wood, believing it to be made up of atoms, we could quite properly say that we had seen and touched a great clump of atoms. If this is not what 'atom' means, what does it mean? If the theory that terms like 'atom' are not descriptive and do not refer to real things has any bite, it must give an account of what such terms do mean.

A theory of the meaning of the word 'atom' might be worked out by comparing its role with that of another well-known term, 'force'. Forces are as unobservable as atoms. Any idea we may have of their presence derives from our observing their supposed effects.

But a little reflection on the role of the term 'force' in classical mechanics shows that it is used to connect up the *descriptions* of the states of systems of things before and after they have interacted. Interaction by collision, or through connecting rods and strings and so on, is recognized in mechanics. The term 'force' is introduced through the action and reaction law, that is by the principle that in every collision, for example, equal and opposite forces act between the colliding bodies. The force-concept can also be introduced in the account of such phenomena as the orbiting of a satellite around a heavy body like the earth. A 'force of gravity' is invented, equal and opposite to a centrifugal force. The two 'forces' enable the relative masses of the two bodies concerned and the velocity and radius of orbiting of the satellite to be connected up in an equation of motion. But when they are so connected the 'force' term has disappeared. It was performing only a formal role. Perhaps 'atom' has the same role in chemistry. Perhaps all the talk about atoms, their properties and their structural organization, is nothing but a picturesque device for connecting up the descriptions of the materials involved before a chemical reaction, and their state after the reaction is completed. It might be argued that nothing factual is added to the description of a reaction when the fact that two volumes of hydrogen combine with one volume of oxygen to give two volumes of water vapour is explained by reference to the atomic formula

$$2H_2 + O_2 = 2H_2O.$$

The observed facts are explained by reference to the hypothesis that the reacting gases are made up of diatomic molecules. Someone who wished to distinguish the role of 'atom' from the role of 'force' in the language of science might insist that 'atom' does have a meaning over and above its role as a connector of facts into systematic, organized knowledge, namely that it means the simplest chemically independent part or piece of a material. Is it right to insist upon its independent meaning? The resolution of this very question will occupy us a good deal in a later chapter.

Concepts

In talking about descriptions and explanations, I have used the word 'concept'. It should not be taken to be a word functioning like the word 'atom', but rather something like the word 'force'. It is used to distinguish the thinking we do from the speaking, writing, drawing, drafting, and modelling we do. To talk of the concept of 'cause' is to talk of the idea of causality, as it is shown in our use of a causal vocabulary, that is in our use of such expressions as 'generate', 'produce', 'make', 'bring about', 'power', and so on, as it is shown in our use of linkages and gear trains to bring about what we want and as it is shown in the use of models and diagrams of mechanisms to explain the processes observed. So talking of concepts allows a general discussion to go on without specifying in exactly what terms the concept is applied: linguistically, concretely, or as evinced in action.

Scientific concepts can be marshalled into two broad groups. There are those like 'mass', 'length', 'charge', 'force', 'red', 'gold', 'mammal', 'momentum', 'valency', etc., which I shall call 'material concepts'. They are property, kind, quality, and substance concepts which can be used in the description of things, materials, and processes. Things and materials have mass, are red, acquire momentum, exhibit valency, and so on. Then there are those which I shall call 'formal concepts', or alternatively 'structural' or 'organizational' concepts. Formal concepts include 'causation', 'existence', 'identity', and the spatial and temporal concepts. To identify some state of a thing, or the presence of a thing at a certain place at a certain time, as a cause, is not to attribute to it any new quality or power which had not previously been ascribed to it, but rather to come to see it as related in a certain way to other states of things which are its effects. Similarly to say that something exists is not to attribute to it a special characteristic lacked by non-existing things. When two things are said to be identical, this does not describe any physical linkage between them. Concepts such as 'existence' and 'identity' impose an intellectual organization upon our observations. They are used to express our understanding of what is happening.

Material and formal concepts can be classified further. For instance, some differences have emerged between the role of the force concept in classical mechanics and the role of the atom concept in classical chemistry. The force concept seems to function in such a way as to make it difficult to treat it as referring to real influences between real things, whereas the atom concept does seem to refer to something over and above the proportions of the reagents.

The atomic hypothesis is not just a picturesque way of expressing the Law of Integral Proportion. The word 'atom' seems to mean the smallest part of the materials with which chemists work which behaves chemically. The question as to whether there are any atoms does seem to be an empirical one, even though it may never be answered in an entirely conclusive way. Other distinctions emerge. The concept of momentum is used for a characteristic of moving bodies that is not observable by our senses, whether naked or extended, whereas redness is an observable characteristic of things. Redness is like smoothness, in being an observable characteristic, but differs from it in being simply correlated with a quantitative concept, namely the wavelength of the light reflected from the thing to the eye when we are seeing something red. Length differs from temperature in that while length is measured by lengthy things, temperature is measured by something other than the internal energy of a material, namely by the expansion of a column of fluid, or by changes in electrical conductivity, or even by changes in colour. We shall be particularly concerned with some of these differences between material concepts when we come to consider, at a later stage of our study, the ideals towards which scientists aim in describing their observations of the properties of things, and of the behaviour of materials.

The aim of philosophy of science is to elucidate the principles assumed in science. These will be found through studies in logic, metaphysics, and epistemology. But the philosophy of science must be related to what scientists actually do, and how they actually think. In the chapters which follow, ideal forms of reasoning, ideals of truth and objectivity, and ideal systems of concepts will be unfolded. Some of the principles that have been proposed in earlier philosophies of science and have been followed by other generations of scientists will be criticized. We shall be 'checking out' philosophical thinking by reference to the actual practice of science, because we must not formulate ideals in our theory of science which are incompatible with any important scientific practice.

Summary of the argument

1. (a) Philosophy as a justification for doing something is distinguished from philosophy as the general theory of some intellectual activity.

(b) This book is concerned only with the first three of the four branches of philosophy, logic, epistemology, metaphysics, and ethics.

2. *Logic* is the investigation of the principles of correct reasoning.

(a) Our method will be to try to find principles which justify intuitions of validity.

(b) It will not be assumed that principles of correct reasoning in one field are adequate to other fields, e.g. the relations of evidential support used in scientific reasoning find no analogue in the principles of correct mathematical reasoning.

(c) Analysis of examples shows that simple notions of validity are inadequate to handle scientific reasoning.

3. *Epistemology* is the theory of knowledge, and is concerned with such problems as the distinction of genuine from spurious knowledge.

(a) (i) As an example of an epistemological investigation, consider the difference between the kind of doubts which might arise over whether the inside of a thing is the way we would expect it to be from an outside view, and the methods of settling them; and the kind of doubt expressed in the feeling that perhaps all life is a dream.

(ii) The point of raising the latter doubt is to explore the limits of application of concepts.

(iii) Such extreme doubts are resolved by showing that in formulating them a word whose meaning depends upon a contrast, e.g. the meaning of 'dreaming' depends upon the meaning of 'waking', is used in such a way as to deny the contrast, e.g. by supposing that all waking might be dreaming.

(b) (i) How do we know that the kinds of processes we can observe here and now are good indicators of the kinds of processes that occur at very distant places, at other times, and among things too small to be observed?

(ii) One resolution is to deny that scientific statements refer to anything other than what can be observed here and now.

(iii) Another resolution is to argue that some instruments extend the senses, and yet others detect influences which cannot be sensed, but which cause sensory experience.

4. *Metaphysics* is the study of the most general concepts used in science and ordinary life, through the study of the internal structure of the language used in different fields.

An example would be the study of the relation between the conceptual system of space and time and the concepts of 'thing' and 'cause'.

5. *The relation between logic and epistemology*

There are very strong connections between the three fields of philosophy in the context of science.

(a) It might be argued as a logical thesis that the relation between evidence and Laws of Nature is such that it should not be construed as a species of inference. From this the epistemological thesis might be extracted that while it is proper to speak of the truth of a piece of evidence it is not proper to speak of the truth of a law of nature.

(b) The epistemological status of predictions might be thought to follow from the logical structure of the making of a prediction, in which initial conditions together with a conjectural law of nature combine in the deduction of a prediction, which shares the conjectural status of the law of nature.

(c) In some cases the law or theory has greater claim to belief than the data, e.g. evolutionary theory.

(d) Scepticism of a radical kind can be generated by demanding that all knowledge be certain knowledge. This can be resolved by treating 'knowing' as a contrast concept to 'guessing'.

6. *The relation between logic and metaphysics*

(a) Logical analyses of general propositions usually employ the concepts of 'predicate', 'variable', and 'quantifier'. The analysis transforms a statement in terms of common nouns into a statement in terms of adjectives. In the process the subject matter might be thought to have changed from things and materials to qualities and processes. Thus one might be led to the metaphysical theory that things and materials should be treated as collocations of qualities.

(b) The connection between the use of common nouns and a metaphysics of things has an influence on basic scientific questions; for instance, the use of words such as 'neutron' and 'electron' in physics tends to suggest that the basic entities are thinglike. The connection between the metaphysics of things and the principles of classical science makes for difficulties in interpreting the new forms of laws in quantum mechanics.

7. *The relation between epistemology and metaphysics*

(a) The epistemological theory that constrains one to deny the accolade of knowledge to propositions about anything other than the phenomena of immediate sensation can be used as a basis for the metaphysical theory that the world of science is the world of sensations, and that the most fundamental existents are sense-data rather than things. This is one of the bases of 'Phenomenalism'.

(b) The metaphysical theory that denies intelligibility to empirical concepts of absolute space and time leads, through the theory of special relativity, to the epistemological theory that whatever may be the actual relations of events in time we can never know which are absolutely simultaneous at different places.

Philosophical principles are an integral part of science, and in the philosophy of science these principles are critically examined. Decisions on such matters as whether to treat thing-concepts as fundamental, as to whether the world is fundamentally atomic, cannot be made by experiment, and are the result of philosophical debate.

8. *The world as it is and as it is perceived*

We know about the world through its effect upon sensitive objects, observers and instruments. The distinction between the way the world seems to be and the way it really is develops in ordinary practical life.

(*a*) Observation and detection:

(i) The world affects sensitive objects such as people either immediately or mediately. There is a tendency to describe these effects in terms of sensations. This, however, is a philosophical theory of small merit.

(ii) The effect of the world upon detecting instruments should be construed as showing the existence of entities which could not become perceptible.

(*b*) The nature and assumptions of experience:

(i) Common experience is not of sensations but of things.

(ii) To treat the world one perceives as a world of things involves metaphysical assumptions, such as the identity and existence of things when not being experienced or detected.

(*c*) Percepts and sensations:

(i) Many features of percepts are not simple sensations.

(ii) Some philosophers believe that these features are an integral part of the experience, but should be treated as given *a priori* by the person having the experience.

(iii) Others believe that only the sensation component of experience is empirical, so that science should be confined to the study of sequences of sensations.

(iv) I shall argue that there is a contribution from the knower to the known, that it is most naturally treated as derivative from the language used to describe the known, and that it is an inescapable but *a priori* metaphysical prescription of reality.

9. Problems that arise from the distinction between the world as it is, and the world as it is manifested in perception.

(*a*) How can we justify the assumption that detecting instruments are detecting entities that are really there, the assumption that there is more to the world as it is than as it seems to be?

(b) Given that there is a contribution from the knower to the known, perhaps true objectivity lies in eliminating that contribution, and dealing only with what is presented in sensation.

(c) What is the justification for treating the detecting instruments as perceivers, rather than just as presenting sequences of phenomena?

The natural sciences resolve these problems by supposing that their explanatory models are representations of the real world, which is not identical with the world as perceived. They acknowledge the role of the scientist in contributing to the scientific picture of reality, and in so doing, acknowledge the impossibility of absolute objectivity in science.

10. *Thinking and experimenting*

(a) Theories and explanations are the archetypal products of the intellectual activity of scientists.

(b) Theories are expressed in sentences, and supplemented with diagrams and models. The theory of the structure of sentential arrangements is logic; the theory of diagrams, models, and pictures is the theory of iconic models.

(c) A theory, or picture of the workings of nature is prior to any experimental programme. Thus we must resist the temptation to construe science as the inexorable accumulation of attested facts.

(d) Thus a scientific explanation is not just an account of the conditions under which phenomena occur, but must include an account of the means by which, in those conditions, the effects are brought about.

11. *The involvement of theories in description*

(a) Metaphysical theories are involved in the very choice of language in which to express a description. We have already noticed the metaphysical difference between a descriptive language of common nouns and one constructed out of adjectives and quantifiers.

(b) Scientific theories are involved in the common use of metaphorical terms.

12. *Examination of the idea that theories are not descriptive*

(a) It has been argued that theories do not describe reality.

(b) Compare the use of the concept of 'force' with that of 'atom'.

(i) The concept of 'force' is introduced to assist in the creation of a logical structure among statements of observation, and to ensure uniformity of method in mechanics. Thus forces are invented together with equal and opposite forces.

(ii) The concept of 'chemical atom' does perform the organizing and unifying role of the concept of 'force', but it is also used for the smallest constituent of observable samples of material, thus having some claim to empirical and descriptive status.

13. *Concepts*

I shall use the word 'concept' in the usual way, that is to refer to the unities of thought in speaking, writing, operating, and so on.

(*a*) Material concepts can be used in the description of things, properties, and processes, observed or unobserved. They can be further classified as they are used in reference to things known to be real, things which might be real, and fictions.

(*b*) Formal concepts are used to bring structure and organization into our descriptions. Such concepts as those used in the spatio-temporal system, the concepts of causality, existence, and so on are formal concepts in this sense.

2
The Forms of Reasoning in Science

IN LOGIC, as I have already explained, we study the ideal forms of reasoning. We do this by trying to find the rules for valid and sound arguments, rules which can direct our thought along certain desirable lines, such as those involved in the pursuit of truth. What we want to avoid above all else is the drawing of false conclusions from true evidence. The point where one starts in a piece of reasoning will be those statements or principles which one supposes that one knows, or for the moment pretends that one knows in order to see what are their consequences. Usually one wants to start from truth, from statements of fact, or at least from statements that one takes to be pretty certain. Such statements form the starting-point of reasoning and are called the premisses of the argument. By an argument I do not mean a dispute, but rather a series of steps of reasoning, which lead from one's starting-point to a conclusion. And since further steps of reasoning can pass beyond one's initial conclusions to other conclusions, which may be of considerable practical, as well as scientific importance, it is clearly vital that the reasoning of scientists should be as perfect as the nature of the case admits.

Actual science is a very complex activity, so it is not surprising that there have been several theories, expressing different ideals of scientific reasoning, particularly for those steps of reasoning by which laws of nature are formulated on the basis of factual evidence, and by which the effect of new evidence on our confidence in the truth of laws is assessed. Before I set out the various theories that are of importance we shall have a look at two typical examples of scientific reasoning. In one a conclusion is drawn which is agreed to be a law of nature, and the conclusion is based upon evidence already known. In the other a law, which is already known, is tested by drawing conclusions from the law and testing them by further observation. The reasoning that leads to theories is very much more complicated and will occupy us later. We should also remember that the reasoning which leads to the original stating of a law is commonly not as simple as the cases I propose to discuss here.

Mendel's Laws

These laws describe the ratios of the numbers of individuals which in successive generations of adult plants and animals show certain characteristics. For instance, the children of a brown-eyed man and a blue-eyed woman will all have brown eyes, but if one of those children marries someone from a similar background, then the chances are that in every four children of such marriages three will be brown-eyed and one blue-eyed. Mendel formulated general laws which described such cases, where dominant (brown eyes) and recessive (blue eyes) 'genetic factors' are grouped in various combinations in people, animals, and plants. Mendel reached his laws by reasoning from evidence. He and his assistants grew pea plants, and cross-bred them, and counted the numbers having certain specific characteristics in each generation. From this evidence he generalized to reach his laws. In so doing he ignored small discrepancies from the perfect ratios which are expressed in his laws. The characters he investigated were the shape and colour of the peas produced in each generation. He compared round with wrinkled, green with yellow.

This process of reasoning could be set out schematically as follows:

In experiment 1 the ratios were 5,474 round, 1,850 wrinkled = 2.96 : 1.

In experiment 2 the ratios were 428 green, 152 yellow = 2.82 : 1.

These are Mendel's actual figures, though some doubt has recently been cast on their authenticity. Then he formulated his law as follows:

In the second generation the ratio of dominant to recessive characters is 3 : 1.

It is perfectly clear that his process of reasoning begins with the experimental evidence and passes to the law as conclusion. Using the terminology of logic we seem to have premisses, the description of the results of the experiments, and a conclusion, the statement of the law. We thus seem to have an argument, a step or series of steps, from premisses to conclusion. What then, we might ask, is the rule or principle of the argument? In trying to answer this question we shall run into all sorts of trouble, so much trouble in fact, as to incline some people to say that there is not really an argument here at all, only the semblance of one.

Kepler and the orbit of Mars

In *De Stella Martis*, Kepler describes his agonies of mind in trying to work out from the observations of the relative positions of Mars against the background of the fixed stars that he had from Tycho Brahe and others, exactly what must be the shape of its orbit. He was trying to pass from premisses to a conclusion, from the facts ascertained about the positions of the planet to a law or at least a general proposition about its motion. His only conclusion was that it must be an egg-shaped orbit that changed its shape according to a theory of librations, expansion and contraction of its diameter. Kepler knew in his heart that this was no good. Then, after working on the theory of the ellipse in another context, he suddenly had the bright idea of *supposing* that the orbit of Mars was actually elliptical, and seeing if, on this supposition, he could work out positions for the planet which corresponded well with the observed positions. This he did, and to his delight, he found that the supposition, or hypothesis, that the orbit was elliptical, was borne out in practice—that is, Mars was seen to be in the positions he had predicted by assuming its orbit was elliptical. He had formed his law independently and then *tested* it against the evidence. Here we have a different logical pattern. In this process the law becomes the starting-point for the reasoning; the premiss and the steps of thought yield conclusions which are judged by their agreement with the facts. However, there is another step associated with this way of proceeding in science which brings it rather closer to the kind of reasoning we have seen Mendel presumably using. What is the point of the tests? Surely if they turn out to be satisfactory we can say that the supposed law is satisfactory. That is, we seem to be inferring or reasoning from the satisfactoriness of the predictions to the satisfactoriness of the law, or if you like, from the truth of the predictions to the acceptability of the law, and that is something like reasoning from the facts to the law, as we supposed Mendel must have done.

Mill's Canons

The first of our examples exemplifies the *inductive* method, the second exemplifies the *hypothetico–deductive* method. Some logicians have treated these methods together as two different aspects of the same reasoning procedure; others have supposed them to be fundamentally different. Many philosophers have attempted to schematize the patterns of reasoning which lie behind such apparent examples of the use of the inductive method as the discovery of Mendel's laws. It is sometimes mistakenly supposed that Francis Bacon was one of these, and indeed the inductive method is sometimes, with little justice, called the Baconian method. The chief exponents of the inductive method have been William of Ockham, John Herschel, and John Stuart Mill. With less than historic justice the usual formulation of some alleged inductive laws has been called Mill's Canons. Mill, Herschel, and Ockham all agreed on the following principles.[1]

The Canon of Agreement: 'If two or more instances of the phenomenon under investigation have only one circumstance in common, the circumstance in which alone all the instances agree is the cause (or effect) of the given phenomenon.'

The Canon of Difference: 'If an instance in which the phenomenon under investigation occurs, and an instance in which it does not occur have every circumstance in common save one, that one occurring only in the former, the circumstance in which alone the two instances differ is the effect, or the cause, or an indispensable part of the cause of the phenomenon.'

In comment upon these Mill says: 'The Method of Agreement

---

[1] J. S. Mill, *A System of Logic* (London, 1879), Bk. iii, ch. 8.

stands on the ground that whatever can be eliminated is not connected with the phenomenon by any law. The Method of Difference has for its foundation, that whatever cannot be eliminated is connected with the phenomenon by a law.' Mill also quotes various other canons of lesser importance. These two main canons he offers as the principles or among the principles of inductive reasoning, for, having found the cause we have found the law, or so he thinks. Here are principles by which we can pass from facts to general laws.

How do these apply in practice? Consider our example of Mendel's reasoning to his law. The only difference between smooth green peas and wrinkled yellow peas which were cultivated in his garden must lie in differences between the parent plants. So whatever differences there were between the parents must be responsible for the differences in the offspring. And looked at from the point of view of the Canon of Difference the agreement in other cases between parental differences and differences among offspring gives powerful support to the detailed hypotheses about how the differences in the generations were to be explained.

But there are difficulties with Mill's Canons, and these I must now gradually bring to the fore. In trying to apply the Methods of Agreement and Difference it is necessary to form some idea of the totality of possible causes for a phenomenon. Suppose we are studying the growth of plants, and we find that in warm weather plants grow more vigorously than in cold. Unless we realize that in warm weather the sun's *light* shines for a longer time we might be tempted to suppose that it is the difference in heat which is responsible for the different rate of plant growth. In this case the greater warmth accompanies the greater growth because both happen together, though the cause of increased growth is the greater amount of sunlight in the summer. Mill's Canons by themselves, could not possibly allow us to decide the question of whether it was the heat or the light which was the causal factor, *unless a different experiment was carried out*. We could try to grow plants in the absence of light and presence of heat, and in the presence of light but in the absence of the usual summer heat. *Then*, using Mill's Canons, we could reason from the results of the experiments which is the cause of growth. But could we? Even after these additional experiments, the results are equivocal. What we actually find is that a certain amount of heat is necessary to stimulate growth, and also light is needed. But light seems to be the predominating cause. But perhaps light is accompanied by a third factor, which is really the cause but which we haven't yet spotted. To resolve this kind of difficulty something quite different needs to be done. We need to investigate the mechanism of plant growth, the process by which a plant synthesizes new material. And when we do this we find that the process depends upon light. It is photosynthesis. Only after the mechanism of growth has been discovered can we be sure that we have the cause. And to describe the mechanism of growth is to put forward a theory as to how growth occurs. Our belief in this theory will depend on how certain we are that we have uncovered the true mechanism. Whether any particular application of Mill's Canons yields information of value is determined by how good a theory we have to explain the processes we are investigating. What information the use of the Canons yields will depend at least as much upon theory held by the investigator as upon what he observes in his controlled experiment. It looks as if Mill's Canons are, at best, a preliminary to the deeper studies of scientists. In effect they eliminate possibilities, but do not positively prove anything. In practice we never rest content with laws for which there are no explanations.

But there is another difficulty about Mill's Canons, if we regard them as expressing acceptable forms of reasoning. We noticed this problem in the first chapter. It is the problem of how far we can regard laws produced by reasoning according to Mill's Canons or any other inductive method as true. Suppose our only reason for believing Mendel's Laws was the experiments Mendel did. And let us suppose we accept the results he puts forward as facts, that is that we accept these as irrefutable evidence, as the actual results of experiments. Can we be equally sure of the generalized Mendelian Laws? What sort of doubts might we have? Well, is it not possible that the figures he got were a coincidence, and that if we made similar studies another year we would get very different results? There is nothing in the experiments that leads us to think that we would get similar results in another year, or in another century for that matter. And yet enunciating the results as *laws* certainly suggests a strong expectation that inheritance has worked according to this pattern and will always do so. The whole idea of there being Laws of Nature carries with it the suggestion that the patterns in phenomena repeat themselves. Yet, where is the evidence for this?

Suppose we answer that the evidence is the whole body of science as we have it. Over and over again we have found that the patterns of nature we have discerned with the help of Mill's Canons do repeat themselves. Natural processes do continue in the ways they have previously proceeded. So we have good empirical grounds for believing in the uniformity of Nature. It seems that Mill's Canons are a pair of logical principles grounded in fact. And yet what logical principles did we use to reason from the evidence of past successes of science to the truth of the supposition that patterns repeat themselves in similar circumstances? None other than Mill's Canons! Where patterns did repeat themselves science was successful; in those cases where 'laws' were based upon coincidences or faulty experiments they subsequently turned out to have exceptions, or sometimes no application at all. Here are the Methods of Agreement and Difference at work again. It looks as if the proof of one of the most important kinds of scientific reasoning depends upon assuming the correctness of that very method itself. So unless one is very careful one can be surprised into thinking that scientific method is either ungrounded, or based upon a fallacious proof. But this dire consequence can be sidestepped.

What we have actually found out is that Mill's Canons are some among many subordinate and limited forms of reasoning. They set standards of procedure, and like other ideals are not in need of practical scientific proof. We have also discovered that if we accept Mill's Canons as ideal forms of reasoning they do not cover all we would want to include in scientific method. Our reasons for thinking that one kind of phenomenon is the cause of another kind are not just a matter of seeing if the two kinds of phenomena appear together or in a sequence, and the second never without the first, but are based much more upon our knowledge or speculations about the mechanisms by which the two are related, and by which the first kind of phenomenon produces the second. Our knowledge that vibration is a cause of metal fatigue derives not only from the fact that vibrated metals break earlier and more often than non-vibrated ones but also from our knowledge of the structure of metals, and the changes in that structure brought about by vibration. Mill's Canons represent one of the forms of reasoning in use in science, but though they are often used as an essential preliminary stage of an investigation they are certainly not the only principles required to formulate hypotheses successfully.

Inductivism

But in the inimitable way of philosophy the canons have been promoted as a complete theory of science. This theory can be expressed in three principles.

The Principle of Accumulation: that scientific knowledge is a conjunction of well-attested facts, and that such knowledge grows by the addition of further well-attested facts, so that the addition of a new fact to the conjunction leaves all the previous facts unaltered. It is as if chemistry consisted only of list after list of reactions among the elements and compounds.

The Principle of Induction: that there is a form of inference of laws from the accumulated simple facts, so that from true statements describing observations and the results of experiments, true laws may be inferred. Mill's Canons, for instance, might be offered as the principles of such inferences. The Laws of Nature are nothing but the codified and generalized particular facts. As Mach put it 'they are the mnemonic reproduction of facts in thought'. In modern science the operation of this principle is often seen in the effort to obtain numerical data and then find algebraic functions to express them.

The Principle of Instance Confirmation: that our belief in the degree of plausibility of (or our degree of belief in) a law is proportioned to the number of instances that have been observed of the phenomenon described in the law. For instance, the more gases we find to be diatomic (having two atoms in their molecule in the gaseous state) the more ready we are to believe and to accept a law that all gases are diatomic.

This is a very seductive theory of science. It seems to be a hard-headed, straightforward, and empirically based view. Scientists are seen as steadily piling up facts, generalizing them into laws, and piling up more facts, step by step in the laboratory. If you can infer the laws from the accumulated facts, you can deduce the facts again from the laws, and the content of the laws is nothing but the facts.

Objections to inductivism

But inductivism will hardly stand a moment's serious criticism. None of its three principles will do at all. Take the principle that science grows by the accumulation of facts. This is just not true. The growth of science is a leap-frog process of fact accumulation and theoretical advance. A change in theory can turn seeming facts into falsehoods. For instance, consider the history of the determination of the atomic weights. What *were* the facts? Under the influence of Prout's hypothesis some chemists considered that the discrepancies between integral values for the atomic weights of the elements were errors, since Prout had maintained that all elemental atoms were combinations of whole numbers of complete hydrogen atoms, and hence their atomic weights had to be integral numbers by comparison with hydrogen. Those who did not accept or had abandoned Prout's hypothesis were inclined rather to suppose that the non-integral weights were the facts, that is a genuine measure of a natural phenomenon. What the facts were depended in part upon whether one held or did not hold a particular theory.

Not only does a change in theory result in a change of fact but even in the field of a single theory there are problems as to what are the facts. Consider again the atomic weights. Are they the relative weights of atoms, as their title would suggest, or are they a picturesquely named set of numerical ratios of the relative weights in which substances combine? Even though these are important considerations, nevertheless it might be thought that we ought to be able to discern some 'brute facts', that is facts which would remain the same throughout all change of theory and point of view. Unfortunately the attempt to find such facts leads to a fatal dilemma. The only facts which seem to be genuinely independent of any scientific theory are those of the present experiences of touch, taste, smell, hearing, and sight that each individual scientist is currently experiencing. But such facts are not, of course, public facts, they are private to each individual. So we have the dilemma, that if facts are truly independent of theory they are private and do not form part of the public domain of knowledge; if they are public facts they are affected by all sorts of influences particularly from previous knowledge and upon which their exact form and our confidence in them depend. At least for science, there are no brute facts. There are no facts which other facts may not change; there is no knowledge altogether independent of theory.

The Principle of Induction leads us into still deeper water. Not only are there the objections that we have already seen to Mill's Canons, which are particular forms of the inductive principle, but there are more general problems, the most crucial being the indeterminacy of the results of trying to use the principle to infer a law. A principle of inference is no good if, from the premisses offered, more than one mutually incompatible conclusion can be drawn. But from each set of premisses of experimental and observational fact, infinitely many laws can be inferred using the principle of induction. This is perhaps simplest to see in a graph.

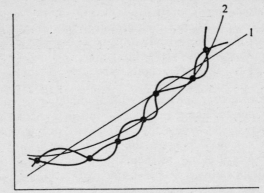

FIG. 1 Curve fitting

Each curve represents a law inferred inductively from the premisses represented by the points marked on the paper, and from each law the points (the facts) can be inferred. Which of these potentially infinitely many laws is the correct one? The principle of induction will not give us an answer, because, from all we can tell using that principle alone, all are equally correct. To deal with this problem inductivists have nearly always added a Principle of Simplicity to their logical armoury. Of all the laws which induction yields, only the simplest should be accepted, they contend. So in the diagram above we should choose (1). But this is a highly dubious principle.

First, it is certainly not clear exactly how to apply it. Should we choose (1) or (2) in the diagram above? Certainly (2) is a pretty good fit, and yet so is (1). It is not difficult to conceive of further experiments adding further points which would remain equivocal with respect to the choice between (1) and (2). Remember that on the inductivist view there is nothing else to go on but the experimental facts, and the principle of simplicity. The step that scientists usually take of referring to theory to adjudicate is no longer open.

Secondly, the history of science can offer little comfort to believers in simplicity. The progress of knowledge does not lead to the most complex conceivable forms of law being discovered, if such a concept as 'the most complex' even makes sense, but it certainly does not lead to the simplest. The growth of knowledge of the behaviour of gases has led from $PV = RT$ to $(P + a/V^2)$ $(V - b) = RT$; the growth of knowledge of the movements of the planets from the simple hypothesis of circles as orbits has led to the more complex curves of elliptical form; the growth of knowledge of the figure of the earth has led from the simple idea of a spherical shape through the idea that it is an oblate spheroid (flattened at the poles) to yet more complex shapes. There can be no doubt that the history of science shows that the laws of nature are always more complex than we originally thought. The Principle of Simplicity as a *blanket* principle can hardly be accepted. Of course at each stage of knowledge it would be mad to choose any more complex hypothesis than one has to, but that it is hardly a methodological principle of the portentous epistemological status assigned to the principle of simplicity.

We have already run into difficulties about confirmation by instances in various contexts, so we can hardly give our allegiance to the Principle of Instance Confirmation. However, the deepest problem of all with inductivism as an exclusive theory of science, offering an all-embracing ideal, is its failure to include explanation in the field of scientific endeavour. This defect can be clearly discerned if we compare Babylonian and Greek methods of astronomy. The practical problem to which scientists from both cultures addressed themselves was the construction of tables of ephemerides, that is tables which would give, in advance, various astronomical facts in which people were interested, such as the times of rising and setting of the constellations, and of the sun and the moon, throughout the year. The Babylonians used an inductive method, the Greeks a non-inductive one. For the Babylonians the tables of ephemerides were constructed by the use of numerical rules, derived as inductive laws, by which addition or subtraction of constants according to definite rules produced sets of numbers which represented the successive risings and settings of those bodies which were of astronomical interest. For details of this interesting kind of astronomy see Neugebauer, *The Exact Sciences in Antiquity*.[1] In this sort of science we have a theory only in the sense of

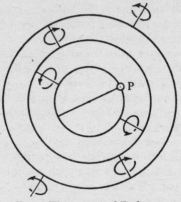

FIG. 2 The system of Eudoxus

an abstract calculating device yielding predictions, but not a theory in the sense of an explanation of the phenomena within some picture of the stellar system, which would account for and explain them. Babylonian astronomical methods, as they have been reconstructed from tablets, could never yield realistic hypotheses about the structure of the Universe and the motions of its parts, since there is no element in the 'theory' capable of realization in the required way. It is just not possible to make a physical hypothesis out of the subtraction or addition of some constant number alone.

The situation in Greek astronomy was quite different. The lunar theory of Eudoxus is already half-way to being a model of the Universe, for its elements are not arithmetical relations but geometrical spheres. For example, to account for the motion of the

[1] O. Neugebauer, *The Exact Sciences in Antiquity* (New York: Harper, 1962), ch. 5.

moon, as we see it, Eudoxus proposed a theory. He considered that there might be (or might be envisaged to be) three spheres, concentric with the earth, each rotating within the other, as in the diagram, the moon being attached to the point P.

Aristotle went further and treated the spheres of Eudoxus as actual physical objects in the Universe, whose real motions actually caused the motion of the heavenly bodies as seen from the earth. Babylonian astronomy gave a 'mnemonic reproduction of the facts in thought', and its method of adding and subtracting (zigzag functions) came as near to a science of purely inductive laws as it is possible to get. But the Babylonians could not explain the observed motions of the stars. On the other hand Greek astronomical theory both summarized the facts and so served as a basis for prediction, and *also* explained these facts to the satisfaction of the proponents of each theory as flowing from the operation of certain mechanisms. It is worth noting, too, that for the Babylonians, the facts were the relative positions of the stars as we see them in the heavens. But for the Greeks the facts were subtly different. What most of the Greek astronomers thought they were seeing when they looked up at night were bright objects which moved about in some pattern or system which was not exactly the way it looked from the earth, because we see the stellar system from only one point of view. It is Greek explanatory astronomy that developed into modern ideas about the Universe, not Babylonian inductive astronomy.

Our conclusion must be that though Mill's Canons are valuable schemata for organizing an investigation they could hardly be generalized into the whole of scientific method. And the reason, in short, is that scientists are not exclusively concerned to discover correlations among phenomena, but are at least as interested in the explanations as to why the correlations that can be discovered are the way they are, and in explanations as to why there are the structures that there are. Astronomy should explain not only the correlation between the rising and setting times of heavenly bodies but also the way the planets are arranged in the solar system, and that system in the galaxy.

This leads to a point of more general application. The real reason why inductivism is so wrong is that it is so unrealistic. It is an attempt to codify a more or less mythical conception of science. It is quite clear from Mill's Canons that inductivists picture scientists experimenting in the *hope* of finding correlations among phenomena. What then are we to make of the activities of anatomists who investigate the structure of organisms, of crystallographers and nuclear physicists who also are searchers after knowledge of structure? What are we to make of the enormous amount of effort put into attempts to measure natural constants accurately? What about taxonomists and their efforts to classify the plants, animals, and minerals found in the world? It would be quite ridiculous to rule out these activities as non-science. And if they are science and, as I contend, a major part of it, science can hardly be pictured as the search for correlations among types of events. Chemists are interested in finding what materials they can make from what; they are not interested in the correlation of the events of the vanishing of one lot of substances and the appearance of another lot. This myth of phenomena has been much in the minds of philosophers and has been responsible in no small part for the pointlessness of much philosophy of science in the past.

Popper's Doctrine of Falsifiability

If science is not a process of the confirmation by instances of inductively derived laws of nature, perhaps it is a process of falsification by instances of conjectures which have no logical basis in the facts hitherto known. This theory, 'fallibilism', has been developed in modern times by Sir Karl Popper.[1] This view retains the inductivist picture of laws of nature as general statements of correlations among phenomena, and evidence as particular statements of correlations among observables. This aspect of the theory can fairly be called 'conservative' and I shall refer to it as the Conservative Principle. Popper's more radical proposal is that evidence is valuable *only in so far* as it would tend to *falsify* general statements. A piece of evidence that would usually be regarded as favourable and supporting a hypothesis is treated by Popper, at least in some of his work, rather as an instance of the failure of an attempt to falsify the hypothesis under test. I shall refer to this view of evidence as the Radical Principle. Popper also uses the Radical Principle, in reverse, as it were, to classify hypotheses as either empirical or non-empirical and non-scientific.

[1] See his *Logic of Scientific Discovery* (London: Hutchinson, 1959) and *Conjectures and Refutations* (London: Routledge and Kegan Paul, 1963).

Those which could be falsified by empirical (experimental or observational) evidence he regards as empirical or scientific, and those which could not be falsified by any kind of test he regards as non-empirical or, as he sometimes puts it, 'metaphysical'. I shall refer to this view as the Radical Demarcation Principle: 'radical' because it excludes from science a very great deal of what is usually thought to be characteristically scientific.

There are various interesting consequences of Popper's principles. From the Conservative Principle it follows that the canons of right reasoning, that is the principles of logic, must be restricted to those first set out by Aristotle. The Aristotelian principles of reasoning relevant in this context can be summed up in the traditional Square of Opposition. General statements can be either affirmative, 'All metals are conductors', or negative, 'No metals are conductors'. Either could be a law of nature. The results of experiments could be expressed as 'Some metals are conductors' when the experiments have been successful, or 'Some metals are not conductors' when the attempt to test some metals for conductivity failed. The relations between these four statements can be shown in a diagram:

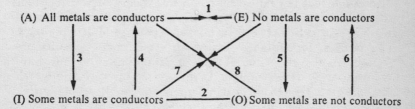

The relations are:

1 A and E cannot be true together.
2 I and O cannot be false together.
3 and 5 If A is true then so is I. If E is true then so is O.
4 and 6 The truth of I supports A, and of O supports E.
7 and 8 The truth of I contradicts E, and the truth of O contradicts A.

Thus 7 and 8 are the relations that contrary evidence has to hypotheses, while 4 and 6 are the weaker relations that favourable evidence has to hypotheses.

The most important consequence of Popper's restriction of the logic of science to **3**, **5**, **7**, and **8** is the view that the process of the generation of hypotheses is not one which proceeds according to critical rational principles. Science has its origin, according to Popper, in a cloud of conjectures about how things go on. This body of hypotheses is progressively whittled down by the work of experimentalists whose results falsify certain of its components. The progress of anatomy, on this view, is to be understood not as the development of correct ideas about the organs and structures of organisms by dissection and observation but as the falsification of incorrect conjectures as to what lies under the skin. The discovery of the capillaries by Malpighi is to be understood not as the confirmation of one of Harvey's hypotheses but as the falsification of some other contradictory conjecture. Similarly, reasoning by analogy, in the view of Popper, has no advantage over guess-work, since in either case the only rational control of hypothesis is the operation of negative evidence. This view of evidence I have called the Radical Principle. It demands a radical reinterpretation of most scientific work, since all evidence which was supposed to be favourable to a hypothesis is now to be treated as an attempt at falsification that has failed. Since on this theory logical relations exist only between falsificatory evidence and hypotheses, no logical relations can exist between favourable evidence and hypotheses. The process, for instance, by which experimental results are averaged to give the value of some natural constant would have to be radically reinterpreted. In the usual interpretation each experiment is treated as bearing upon and finally contributing to the value we eventually assign to the constant. It is that idea that makes sense of including all the results in an averaging. Since experiments yield only potentially falsificatory evidence for Popper, averaging is an irrational process. The experimental results, all of which differ from the final assigned value, would have to be regarded as falsifying the hypothesis that the value was the average value. There can be no notion of acceptable error on this theory, so the usual move of treating divergent results as failing to falsify the hypothesis because they are within the acceptable margin of error, is not open. Hence, so far as I can see, no value could be assigned to a natural constant by this method. Processes such as averaging would have to be expunged from science.

Two vitally important consequences follow from the Radical Demarcation Principle. Two important classes of empirical statements, which are the very type of empirical statements for many thinkers, are to be demoted from scientific status. These are general existential statements like 'There are molecules' and 'There are okapis', and particular probability statements like 'The probability that this die will come up 6 at the next throw is 1 : 6.' The former are non-scientific because they are unfalsifiable, though they are confirmable; the latter are non-scientific because the failure of the dice to come up 6 in that throw or even in the next six throws does not count as evidence strong enough to falsify the statement. Particular probability statements are incorrigible. If in the next 100 throws 6 does not come up we do not regard the probability statement as false but argue that there must be something wrong with the die. The conclusion that general existential statements are non-scientific is made all the more striking by the fact that Popper and his followers hold that evidential statements, expressing the results of experiment, are to be expressed as particular existential statements. For example, 'This is a meter reading of 30°' and 'This is a radio-star' are of that form. Now it follows *logically* that if 'This is a radio-star' is true, then 'There are radio-stars' is true. Does it not seem strange that the former should be classified as empirical and the second should not be so classified?

This theory, for all its superficial attractiveness and its great simplicity, has drawbacks as a comprehensive account of scientific method. First, it confuses psychological conditions for science with logical conditions. It is of course the mark of a scientist that he should be ready to abandon any theory or law if evidence accumulates heavily against it, though, of course, one must also be aware that much fruitful science arises from the effort to explain away contrary evidence and to preserve a law or theory. In fact neither canons of acceptance nor canons of rejection of laws and hypotheses and theories are so simple as to be expressible in the rather gross relations of Aristotelian logic. The joint method of confirmation and falsification has been a commonplace of scientific method since at least the sixteenth century. And it has also been pointed out in fact with particular force in the sixteenth century, that experimental evidence *alone* is not enough either to confirm or to refute a theory or hypothesis, and that other rational procedures of decision must be looked for. Furthermore, in many cases, the proper move to make when faced with contrary evidence to some hypothesis is to reinterpret it as favourable evidence by developing the relevant theory, so that the awkward facts are now seen as being in accordance with the theory. The conditions under which this is the right way to proceed need to be studied carefully, and distinguished from those rather rare cases in which unfavourable evidence would call for the rejection of the theory or hypothesis.

But the conflict between Popper's theory and scientific practice is most acute in the matter of its classification of existential statements. It is just because of their strong logical connection with evidence that they play so great a role in science: indeed, it is in existential statements that most of the permanent advances of science are preserved, against a background of shifting hypo-theses. Knowledge of the existence of chemical atoms, bacteria, genes, subatomic particles, and so on constitutes the hard core of scientific knowledge, while our ideas as to the constitutions and capabilities of these entities develop and grow and change. Existential statements express the permanent empirical part of science. Their study is a subtle one, and not all the entities mentioned above have the same status; but to deny existential statements a place in the community of scientific truths must surely be mistaken.

Finally the most disturbing feature of this theory of science is the consequence that the intellectual processes by which hypo-theses are discovered and theories formulated are capable only of psychological studies, that is, that there is no *rational* process theses are discovered and theories formulated are capable only of phychological studies, that is, that there is no *rational* process of discovery in science. Of course the simple ideas of inductivists as to the rational principles of scientific discovery are pretty silly, but it does not follow from that that there is not a rationale of intellectual discovery and creation. At this point I shall offer only a glimpse ahead, but we shall see that profound and subtle extensions of traditional logic, particularly in the fields of models and analogies, can provide a rationale of creation and discovery, and provide it in just such a way as accords with scientific practice. Falsifiability does not, then, provide a unique ideal of reason.

Positivism

Having rejected the idea that the only role of theories is to provide conjectures to be falsified, we must consider a very closely related theory to that of Popper. Indeed, it may be called the sibling theory. This is an aspect of positivism, indeed its logical aspect. Like Popper, positivists conceive of theories as organized only according to the canons of deductive logic, the logic of mathematics and taxonomy. The effect of this is to force them to conceive very narrowly of theory and its ideal logical structure. It is difficult to set out the logical aspects of positivism without referring to its epistemological and metaphysical sides too, and this I shall be obliged briefly to do.

The main ideal of reason for a positivist is that a theory should be a deductive structure, such as is exemplified in geometry, and there should be the very same structure in a scientific explanation as in a theory from any realm of human knowledge. The idea that science and mathematics should share the same ideal of reason is not a new one, but it has been given a new fillip in recent times by the great developments of the logic of mathematics that have taken place lately.

The view that a deductive structure is the ideal of reason in theory has not always been associated with the positivist point of view. It was held, for instance, by Descartes. The view has two aspects, a picture of theories and a picture of explanation. Suppose there are three laws L_1, L_2, L_3, known by experiment to hold in a certain field of phenomena. A theory is devised to explain them. It follows from the deductivist, positivist view of a theory that a theory has been successfully constructed when some hypothesis H, or some hypotheses H_1, H_2, etc., have been thought of, from which L_1, L_2, and L_3 can be deduced. The hypotheses function like the axioms and postulates of a mathematical system, and the laws like theorems. Kinematics would provide an example of this. From the hypothesis (postulate) that the acceleration of a body be uniform

$$H \qquad\qquad \frac{d^2s}{dt^2} = a$$

various laws of motion can be deduced:

$$L_1 \qquad\qquad \frac{ds}{dt} = at + k$$

$$L_2 \qquad\qquad s = \frac{at^2}{2} + kt + j$$

This has a deductive structure, in that H implies L_1 and L_1 implies L_2.

If the sole aim of theory is to provide a basis for the deduction of laws then some important consequences follow. In many, if not in all theories, there are important theoretical concepts. For example, in the theory of inheritance there is the concept of 'gene' or 'genetic factor'; in chemistry of 'atom', 'valency', etc.; in dynamics of 'force'. The theory which explains Mendel's Laws of the statistics of the distribution of characteristics from parents to offspring makes use of the conception of a genetic factor, and of dominant and recessive genes. From this theory Mendel's Laws can be derived since the laws which govern the behaviour of genes are formally analogous to Mendel's Laws. For many positivists the achievement of the deductive relationship is all that is required. What then is the status of the gene and its dominance and recessiveness? Such conceptions, viewed from the standpoint of deduction, become nothing more than cyphers within the logical system serving a formal role in binding together the laws into that system. To ask for the empirical meaning of 'gene' as an independent entity would be a mistake, for its role is purely logical. Any other meaning it may have simply helps poor logical thinkers to conceive of the logical relations in the theory.

Parallel to this is the deductivist view of explanation. It is enough, deductivists aver, to explain a happening, a fact, or anything whatever, if a description of that happening can be deduced from a 'covering' law. Now what is a covering law? I can best explain by an example. Suppose that the fact to be explained is that copper conducts electricity. The explanation, according to this theory, is achieved by deducing that copper conducts electricity from the proposition that all metals conduct electricity together with the fact that copper is a metal. From these two statements the statement that copper conducts electricity can be deduced. The general statement 'All metals conduct electricity' is the covering law. Similarly if the happening is described as 'This piece of metal conducts electricity', and we know that this is a piece of copper, the happening is explained by asserting that copper conducts electricity, and this then becomes the covering law.

There are many important consequences of these views. Two of great importance follow directly from the deductivist view of theories. Since it is possible that infinitely many theories might be invented from which the same set of laws follows deductively, by adding any old additional terms, there must be some way of distinguishing between good and bad theories other than merely that they imply the laws. For example if the law is

All metals conduct electricity,

then one theory could be

All materials which have free electrons are conductors,
All metals have free electrons,

   therefore

All metals conduct electricity.

But another theory would be

All wooden things are conductors,
All metals are wooden,

   therefore

All metals are conductors.

From a deductive point of view both theories imply the law equally validly! To say that the first theory is true and the second false is to introduce a consideration over and above the achievement of deductive connection. In these circumstances deductivists fall back upon another famous principle, and they advise us to choose the simpler theory. But it is not so easy always to tell which is the simpler theory. Nor is it clear that simplicity has the merit of being a sign of truth. But even without going into that problem again it should be quite clear that as an ideal of reason mere deducibility is equivocal and unsatisfactory considered alone.

The second important point to be noticed is that this theory implies the symmetry of prediction and explanation. It is certainly true that prediction of at least some new laws and certainly of all new particular facts is by deduction. To predict that curium will be a conductor of electricity all we need do is note that curium is a metal and that all metals are conductors, and we can predict by logical deduction that curium will conduct electricity. On the deductivist view the same structure pertains to explanation too. So prediction and explanation are really the same logical process, and differ only in so far as what is predicted has yet to happen while what is explained has already happened. This would imply too that every prediction counts as an explanation after the event, and every explanation counts as a prediction before the event. Neither of these two principles is true. Consider the course of a disease. Long before any explanation of what happens is available the empirical knowledge of experienced doctors may enable them to foretell the course of disease with great accuracy from the symptoms. We would hardly call the description of the symptoms the explanation of the later stages of the disease. Nor indeed would we call the predictions made from nautical almanacs the explanation of the risings, settings, and conjunctions of the heavenly bodies. Characteristically, giving an explanation involves describing the mechanism, usually the causal mechanism, responsible for a series of happenings, and this may not be enough to predict just what will happen. We know the causal mechanism of evolutionary change pretty well, but until we actually observe what happens we are unable to predict the appearance of new forms of plants and animals, because of the presence of the random (unpredictable) element of mutation in the system. Explanation is always possible but prediction is not. It hardly seems right to defend the deductivist position by saying that prediction would have been possible had the particular mutation been known. It was not known, and it could not be known until it happened.

Deductive inference of the laws from axioms and postulates is too easily achieved, if we consider that alone as a desideratum. The axioms and postulates must have some better claim to our credence and attention than *merely* implying the laws. But any attempt to specify that claim to attention leads away from deductivism, because the meaning and empirical content of the hypotheses of the theories becomes paramount over their logical powers.

The ideal forms of reasoning which we have looked at in this chapter are unsatisfactory only if they are offered as exclusive and complete accounts of the logical, rational part of scientific method. Mill's Canons and the inductive method do provide a reasonable preliminary procedure by which hypotheses as to what are the laws of phenomena can be formulated. They do not provide a complete method because they provide neither a test for the hypotheses nor an explanation of the facts. In the end the decision as to what the laws really are depends heavily upon which of the possible hypotheses seems to be connected with a mechanism that could produce something like the pattern it describes. The Laws of Chemical Combination are related to the theory that the mechanism of chemical reactions is a rearrangement of the atoms occurring in the molecules of the reagents interacting. We reason like this: if this is the mechanism of chemical reaction then the patterns of behaviour of chemically active substances must conform to this or that law, the Law of Definite Proportions, for example.

To adopt, instead of the principles of inductivism, the principle of falsifiability is an advance in that it directs our attention to the occasions upon which evidence has a negative function, but it will not do as a complete account of scientific method. It leaves the process of scientific discovery in the dark, and keeps it at best a topic for investigation by psychologists and sociologists, as if there were no canons of reasoning by which rational methods of arriving at theories and better hypotheses could be controlled. Deductivism too, we have seen, promotes a partial and hence unrealistic picture of science, if taken as an exclusive canon of theory. As we shall see when we come to spell out the rational process of the invention of theories in detail, there are ideal forms of reasoning at work in that area of human thinking too. They have to do with the canons of constructing and imagining models, and thus depend upon principles governing the rational way to make comparisons, to judge likenesses against unlikenesses. They lead to ideas of structure more complex than the deductive relationships that are to be found at work in the organized parts of mathematics.

Summary of the argument

1. Logic is the attempt to specify the rules of correct reasoning, where reasoning is typically a passage of thought from some given or assumed statements to others. The complexity of science is evidenced by the number and diversity of ideal forms of reasoning that have been advocated.

2. *Examples of reasoning from the history of science*

(*a*) Mendel's Laws

As officially set out by Mendel the general statement of the ratios of dominant to recessive characters is inferred from the set of ratios found in a number of experiments, by rounding off to the nearest whole number. There is some reason to think, however, that the process of reasoning on Mendel's part was more elaborate, involving a stage such as that described, followed by an application of the hypothetical law to the experimental results to eliminate aberrant cases.

(*b*) Kepler's orbit of Mars

Kepler, after repeated failures to infer a 'good-looking' law from the observed positions, supposed that the orbit was elliptical and used the observations to test the law.

3. Case (*a*) has been called an example of the 'inductive' method, and case (*b*) of the 'hypothetical-deductive' method. It should be remembered that the kind of 'induction' advocated by Francis Bacon is not like case (*a*).

4. *Mill's Canons*

(*a*) The classical exposition of the inductive method is as Mill's canons, the two most important of which are:

The Canon of Agreement, which states that whatever there is in common between the antecedent conditions of a phenomenon can be supposed to be the cause or related to the cause of the phenomenon.

In practice this must be supplemented by the Canon of Difference, which states that differences in the conditions under which an effect occurs and those under which it does not must be the cause or related to the cause of that effect.

(*b*) These canons do express a kind of method, but

(i) they depend upon the assumption that we know the totality of possible causes for a phenomenon.

(ii) In the case where the revealed causal factor is complex, innumerable further experiments are required to distinguish between contributing causes and side-effects.

(iii) The difficulties of (i) and (ii) can be resolved only by adversion to a theory describing the causal mechanism involved.

(iv) The causal laws inferred are not thereby known to be true. Any attempt to establish the truth of a statement inferred from particular facts by their help involves the assessment of historical evidence as to their efficacy in the past. The use of this evidence to provide a general justification of the Canons must invoke the Canons.

(v) We resolve this famous difficulty by admitting that Mill's Canons are only some among the forms of reasoning, and that they are not capable of proof. They specify the preliminary stages of science.

5. *Inductivism*

The idea that something like Mill's Canons exhausts the possibilities of scientific method has been made into a complete philosophy of science, inductivism. Its principles are as follows:

(*a*) The Principle of Accumulation: that scientific knowledge grows additively by the discovery of independent facts.

(*b*) The Principle of Induction: that there exists a form of inference by which laws can be inferred from particular facts, unequivocally.

(*c*) The Principle of Instance Confirmation: that our degree of belief in the truth of law is proportioned to the number of favourable instances of the law.

6. *Objections to inductivism*

(*a*) Theories determine, in part, what are the facts, e.g. Prout's hypothesis and discrepancies from integral atomic weights.

(*b*) Data are transformed into facts by expression in terms derived from some theory, and thus stand or fall with the theory, e.g. atomic weights.

(*c*) Infinitely many laws can be inferred from any known form of induction from a set of facts. Inductivists typically invoke a Principle of Simplicity to decide which among the possible laws is correct.

(*d*) Examination of the Principle of Simplicity:

(i) On any index of simplicity there are still infinitely many laws of equal simplicity.

(ii) The history of science shows progressively more complex laws in each field of phenomena.

(*e*) Inductivism does not admit explanatory force to a central place in science, typically treating it as a psychological phenomenon. On inductivist principles it is hard to see how Greek astronomy could be treated as an advance on Babylonian, in which simple inductive laws were used to infer future astronomical events. Greek astronomers produced the same predictions, but used, for the most part, a realistic model of the Universe in their reasoning.

(f) The fundamental objection to inductivism is that it bears very little resemblance to scientific practice.

7. Popper's Doctrine of Falsifiability

(a) Conservative Principle: laws state correlations among phenomena.

(b) Radical Principle: evidence is important only for its power to falsify hypotheses.

(c) Radical Demarcation Principle: only those general statements which are falsifiable are to be classified as genuine scientific statements.

(d) Consequences of adhering to (a), (b), and (c).

(i) The principles of logic are those of the Aristotelian Square of Opposition.

(ii) The generation of hypotheses is not subject to rational principles and is a topic for psychology.

(iii) There are no logical relations between favourable evidence and hypotheses, other than the fact that a piece of favourable evidence represents a failure to falsify a hypothesis, and thus counts as the passing of a test. Notice that strict adherence to these principles makes the extraction of a numerical constant's 'best' value from data by, e.g. averaging, an irrational process.

(iv) It follows from the Radical Demarcation Principle that both general existential statements, and particular probability statements are not part of science. This is paradoxical in that in each case statements of the forbidden classes can be inferred with deductive rigour from statements which are scientific according to the criterion.

(e) Critique of Falsificationism

(i) General existential statements express the permanent part of scientific knowledge.

(ii) Contrary evidence must *accumulate* before a hypothesis is agreed to be false.

(iii) Particular pieces of contrary evidence are a challenge to the theoretician to explain away.

(iv) A rationale of hypothesis formation can be constructed, provided proper weight is given to theories.

8. *Positivism*

(*a*) The positivistic ideal of reason is that a theory should have a deductive structure, the ideal form of reasoning in mathematics, i.e. a theory should have the form of a theorem.

(*b*) The most important consequence of this view is that concepts which appear only in the theory are construed as having meaning only by virtue of their place in the theory, as logical devices.

(*c*) Explanation is achieved by deduction of what is to be explained from a 'covering law', together with ancillary premisses.

(*d*) It was pointed out as long ago as the sixteenth century that strict adherence to (*c*) implies that a false theory is as good an explanation as a true one. It follows that (*c*) must be supplemented by another prin-ciple, usually the Principle of Simplicity, which we have examined in the inductivist context.

(*e*) Prediction and explanation must have the same logical form, if we accept (*a*), (*b*), and (*c*). But some predictive structures have no ex-planatory force, e.g. a disease syndrome; and some explanations have little or no predictive power, e.g. evolutionary explanations.

(*f*) The main failure of the positivist point of view is that its advocacy of a simple deductive structuring of theory, is too easily achieved to serve as a desideratum for explanation, which is only achieved when the terms in the theory refer to the causal mechanisms at work in the production of the phenomena. Inductivism, falsificationism, and posi-tivism each express one of a complex of useful forms of reasoning.

3
Scientific Knowledge

IN THIS CHAPTER we shall be concerned with theories as to the nature and status of scientific knowledge. In this connection, we shall be concerned with theories about the meanings of certain kinds of words appearing in scientific theories. Science is a collection of well-attested theories which explain the patterns and regularities *and* irregularities among carefully studied phenomena. There are two main kinds of investigation that this suggests. We can ask what is the status of the phenomena about which we are supposed to know something, and we can also ask about the subject matter and content of theoretical knowledge. I want to start by describing five examples, each of which might be thought to provide some sort of model scientific investigation, and ask of each certain epistemological questions. Then we shall see how four famous epistemological theories would treat these examples.

Chemical atoms

Two solutions are mixed together, and a white precipitate gathers in the bottom of the test-tube. The remaining liquid is gradually reduced by heating and when it has evaporated another white substance is found. The original samples of substances which were dissolved to form the two solutions had been weighed, and on weighing the amount of the precipitate and the substance left by evaporation, we find the total weight of material before and after the reaction to be approximately the same. For this and a host of similar phenomena we have explanations in the theory of the chemical atom. The smallest independent parts of materials are molecules, and the molecules are made up of groups of atoms arranged in space with characteristic powers of joining, so that, for example, one atom of a certain kind characteristically joins with two of another kind, and so on. The atoms always preserve their relative weights unchanged during chemical reaction. A chemical reaction consists entirely in the rearrangement of the original atoms into new groupings, thus bringing about a change in the substances involved, by changing the molecules that are present. In the classical period of chemistry when these ideas were being introduced, the only techniques of investigation available were qualitative identifications of different substances, partly by their appearance, partly by their capacity for reaction, partly by their origin, and partly by the results of progressively more accurate weighings of the reagent and the products of chemical action. On this basis what is the status of our alleged knowledge of the behaviour and even the existence of chemical atoms? For a historical account of the arguments over chemical atoms and the attempts to replace the concept by something else see Knight, Brock, and Dallas, *The Atomic Debates*.[1]

Light rays

In trying to understand the way images are formed by curved mirrors and lenses, and how what one sees in a plane mirror is related to its position, in trying to understand the way shadows appear, and in a host of other phenomena, it is found very convenient to suppose that something passes from the object through the glass of the lens or mirror, is reflected by the mirrored surface, or bent in its path by the lens, and finally reaches the eye. Whatever it is travels in straight lines. These lines are light rays. They are the basis of geometrical optics. Once they were thought to be the paths of corpuscles. Later they were supposed to be geometrical abstractions from moving wave fronts, and nowadays they retain something of each of these classical conceptions. But they are not seen in nature, they are drawn on paper. What is the status of that part of our knowledge of optics which is the study of the behaviour of light rays?

[1] W. H. Brock (ed.), *The Atomic Debates* (Leicester: University of Leicester Press, 1967).

Heat

In order to make sense of the results of simple calorific experiments, such as those in which a piece of metal of known weight and temperature is put into a known amount of water at a known temperature, and the ensuing temperature found, we need to suppose that the heat lost by the metal is equal to the heat gained by the water, when they have both reached the same temperature. The very possibility of calculating the result of such an experiment depends upon using this conception of heat and its conservation. But what is heat? Does something flow from the hot metal to the tepid water, heating the water? Is the same stuff flowing from the sun to the earth warming us on a sunny day? What is the status of our alleged knowledge of *heat*, as contrasted with our knowledge of relative temperatures, weights, and quantities? If we look carefully at the way our concept of heat functions in elementary calorimetry and much else besides, we can see that when we come to actually measuring anything, it is always temperature, mass, or volume that is measured, never quantity of heat. What we call 'quantity of heat' is always the result of a calculation. Heat, as it were, cancels out in very many cases. Though not in all, because there are not only equivalences between heat and heat but also between heat and motion, heat and electricity, etc. In those cases heat cannot, at least at first sight, be so neatly eliminated from the equation.

Mechanical force

The concept of force seems to play a very similar role in mechanics to that of heat in calorimetry. Whenever it appears it can be eliminated in terms of less mysterious concepts, usually either strain or mass acceleration. Newton's Third Law enshrines this eliminability. Action always equals reaction, and each can be expressed in terms of masses and changes in motion: there is never any net force which would manifest itself in other ways. But unlike heat, force has no other relationships outside mechanics which would give it some kind of independent status. At least as a concept force is strictly eliminable from mechanics. How should we express the status of our knowledge of forces? We shall see that the case of force has tended to be treated as a typical case, and in some epistemologies other concepts like heat, atom, and light ray have been treated on the model of force, and seen as strictly eliminable from science.

Virus

The idea that disease was caused by the invasion of the body by hostile organisms originated in the seventeenth century and gradually gained ground, to be accepted by practically all medical practitioners as well as research workers by the end of the nineteenth century. However, there were some diseases for which the bacterial agent could not be isolated or identified. Nevertheless the infection theory implied that these diseases too did have agents, but it was thought that they were too small to be seen with the help of microscopes or isolated by filtration. They were called 'viruses'. Were they exactly like bacteria only smaller, or did they act and live according to different biological principles? It gradually came to be realized that they were not really independent organisms. They lived by taking over some of the vital functions of the cells of their host, and directing these according to their own needs. This idea, too, had been introduced in the seventeenth century by Van Helmont with his theory of *arche*, but was subsequently much neglected. Both conjectures then, one as to the existence of viruses and the other as to their mode of living, have been shown to be true. What is the epistemological status of our knowledge of the causes of disease throughout this story? This example seems in some ways to run counter to the force example, and perhaps to provide a different paradigm for the ideal structure for a theory.

The planetary hypotheses

In the sixteenth century, a great epistemological controversy raged concerning the status of hypotheses about the way the planets, the sun and the moon, and the stars moved, and what was their arrangement in the heavens. This will provide us with a very instructive example, and to understand it we must run over the various hypotheses that had been held in the matter.

We have already seen Babylonian astronomy operating with mathematical hypotheses only, simply linking the phenomena one to another without introducing any further concepts other than purely mathematical ones.

The geocentric theories

In the theory of Eudoxus, the earth was at the centre of a system of nests of spheres, there being three or four spheres for each planet. They were connected in such a way that each ran on an axis inside the next (see p. 46 for a diagram of such a nest of spheres). The rotations and counter-rotations at different inclinations of the spheres within these nests caused the motion of the attached planet as seen from the earth. It explained the chief peculiarity of the wandering motion of the planets, that when considered against the background of the fixed stars they halted, and looped, and moved forward again on their course. The daily

FIG. 3 Retrograde motion of a planet

rotation of the heavens was brought about by the rotation of an outer sphere to which the fixed stars belonged. The Eudoxus system conceived of the spheres as concentric, and was treated as a realistic physical theory by Aristotle, though some scholars

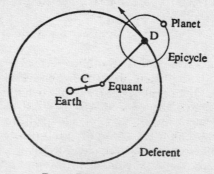

FIG. 4 The system of Ptolemy

are inclined to think that Eudoxus meant it as perhaps no more than a geometrical analysis.

In the theory of Ptolemy, the spheres were replaced by circles, with a complex geometrical relationship. The deferent circle had its centre displaced from the centre of the Universe, and the earth, too, was slightly displaced from that point. The epicycle turning back upon the motion of the deferent explained the peculiar manner in which the planets moved. Each planet had its own system of eccentric, deferent, and epicycle.

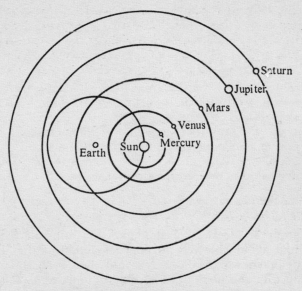

FIG. 5 The system of Ursus and Tycho Brahe

The theory of Ursus and Tycho Brahe involved a system of circular orbits for the planets about the sun, and the whole sun-centred structure rotated about the earth. Such a conception as this was perhaps possible only in the absence of a system of dynamical concepts in which its physical plausibility could be tested. It is significant that it arose in the interregnum between the downfall of Aristotelian physics and the rise of the modern notions of dynamics.

The heliocentric theories

There were two chief theories. Those due to Copernicus used a complex series of circular motions for each planet, including a motion of the earth about the sun. That of Kepler conceived the orbits to be elliptical with the sun at one focus. Kepler's theory

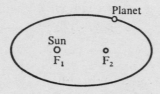

FIG. 6 The ellipse of Kepler

was embedded in some very peculiar ideas about the harmonious structure of the world which, though fascinating, must not detain us here. For a dramatic account of Kepler's theory, see A. Koestler, *The Watershed*.[1] I shall try to explain the differences between the various epistemological theories in terms of these and the foregoing examples. Each of these views amounts to an attempt to define the ideal form of scientific knowledge.

Complete phenomenalism

This group of theories depends upon the principle that only propositions about observed phenomena should be dignified by the status of knowledge. Science would then properly concern itself only with the identification, classification, and relation of phenomena. The different theories that have been held by complete phenomenalists really depend upon differences of opinion as to what are to be treated as the ultimate phenomena. By confining ourselves to phenomena it is hoped that we achieve certainty, and so gain permanent knowledge. It is as if the facts could be identified once and for all and provide a permanent store of knowledge, while theory would be a temporary and eliminable feature of psychological value only, a mnemonic by which the facts were stored and by which they were recollected.

Broadly speaking, there are two kinds of phenomenalist theories,

[1] A. Koestler, *The Watershed* (London: Hutchinson, 1961).

depending upon whether the phenomena which are to be the only true and real subject matter of science are taken to be the ordinary things as perceived, with their ordinary qualities and relations, or whether our perceptions should themselves be further analysed, perhaps into the elementary sensations upon which they are alleged to be based. I shall describe and criticize three important theories under each heading.

The phenomena are what we ordinarily perceive
Patricius

Patricius was a well-known writer on science and its methods towards the end of the sixteenth century. His phenomenalism was prompted and provoked by the confusion in astronomy, where a wide variety of theories were in vogue (cf. pp. 66f). At the time, there seemed to be little chance of distinguishing between them objectively. In his *Pancosmia* of 1591 he advocated the idea that for scientific purposes we should cease to make the distinction between what the planets really do, and what they seem to do. If God could create a Universe of epicycles and eccentrics, Patricius asks, could he not equally well have set up a Universe in which the planets moved through the heavens, and looped back upon their tracks as they seemed to do? In Chapter 18 of the *Pancosmia* he says, 'The planets are called "wanderers" because they stray either side of the ecliptic, and move at varying speeds, now forwards, now backwards now stationary. But although they seem to stray, *in fact they do not at all.* For they are carrying out the commands of the Creator . . .' And he goes on to say that 'as for their relative distances, positions and courses, all the disagreement about this springs from the fact that astronomers have supposed the planets, like the fixed stars, to be carried around fixed to "spheres". What is our word in the matter? Trust the evidence of your eyes . . . They tell you *nothing* about spheres, therefore they do not exist' (translation by T. I. M. Beardsworth). Ramus in his *Scholae Mathematicae* of 1569 expresses a similar view in which he notes that astronomy was done without hypotheses, before the ideas of Eudoxus gained currency, and was none the worse for it.

Berkeley

A more elaborate and sophisticated version of this theory of science was proposed by George Berkeley, particularly in his *Principles* of 1710.[1] His interest in the methodology of science was roused by two aspects of Newtonian science that he found repugnant. He was suspicious of the epistemological status of the force concepts like attraction and repulsion, and equally dubious about the status in knowledge of Newton's concepts of absolute space and time. The theory of science that he advocated eliminated both force and absolute space and time. It came from his general philosophical position. This is a most important philosophical standpoint so I shall try to explain it in some detail.

Berkeley's philosophy takes its start from a denial of the notion that our perception of an object is something different from the object itself, perhaps just an idea excited in us by an object we do not really see. He begins with the common-sense principle that the chair I see is the real chair, and its qualities are real qualities. There is no mysterious material chair which causes my perceptions but is different from them. To exist is to be perceived or perceivable, and that which cannot be perceived under any circumstances cannot be said to exist, with the exception (for Berkeley) of God. In general, things in the world are God's perceptions, and so great is his power that we, too, perceive. It is not that God thinks of a chair which comes into being and then causes us to see it, but that God's chair-thoughts are so powerful that they cause our chair-perceptions. Such perceptions will be orderly and in a rational sequence because God has ideas in an orderly way. Science is the attempt to read God's mind, by discerning the orderly pattern of sequences of perceptions in our experience. To some part of this theory even the most sceptical might assent, namely that part of the theory that asserts that there is no hidden cause of our perceptions. Real things are the things we perceive. But having denied the difference between what is perceived and what really exists Berkeley passes on to another, and for philosophy a crucial step. The particular theory which he is denying is that of Locke, in which a sharp distinction is drawn between objects, things which have the power to affect us, and ideas, sensations, and perceptions which are caused in us by the

[1] G. Berkeley, *Works*, ed. T. E. Jessup (London: Nelson, 1949), Vol. ii.

action of things. Ideas, which for Locke included sensations, and perceptions, which are treated as groups of sensations, are mental, or 'in the mind'. Now Berkeley, gaining assent to his denial of the *difference* between ideas and real things by a reference to common sense, then chooses the surprising and non-common-sensical alternative out of ideas versus real things, to say what these entities are. Chairs, tables, stars, and so on become for Berkeley not only real as perceived, but are themselves ideas. This move gives powerful metaphysical overtones to his philosophy, and is one of the ways in which he makes the introduction of God into his system appear necessary. All causes become of a kind with mental causes, and all agents are spirits. The source of our ideas is then either ourselves or God. Only of God's ideas can we have any reason for belief in an orderly progression, so orderly progressive ideas which *we* would say come from a fairly orderly and coherent world, Berkeley would say come from God.

But Berkeley's philosophy of science comes directly from his epistemology, and can be understood without the divine metaphysical superstructure. Our experiences of happenings among real things are, for him, just experiences of the succession of ideas, and not of agents acting causally. These two points are put by Berkeley in Section cvii of the *Principles* as follows:

It is plain philosophers amuse themselves in vain, when they inquire for any natural efficient cause distinct from a mind or spirit ... from what hath been premised no reason can be drawn, why the *history of nature should* not still *be studied*, and observations and experiments made, which, that they are of use to mankind, and enable us to draw any general conclusions, is not the result of any *immutable habitudes*, or relations between things themselves, but only of God's goodness and kindness to men in the administration of the world ... by a diligent observation of the phenomena within our view, we may discover *the general laws of nature, and from them deduce the other phenomena*, I do not say *demonstrate*; for all deductions of that kind depend on a supposition that the Author of nature always operates uniformly, and in a constant observance of those rules we take for principles, *which we cannot evidently know*.

From all this, Berkeley draws some further conclusions, very characteristic of a phenomenalist point of view. Not only does he argue that we can know only the relations of ideas ('... knowledge of the phenomena ... consists, not in an exacter know-ledge of the *efficient causes* that produce them ... but only in a *greater largeness of comprehension*' ... Section cv). He argues too that the very meaning of theoretical terms which seem to refer to something other than the bare phenomena really have a pheno-menal meaning. In Section ciii Berkeley discusses the meaning of the two key Newtonian terms, repulsion and attraction. 'Is it that the word "attraction" signifies the manner or the tendency?' 'But ... I do not perceive that any thing is signified besides the *effect* itself: for as to the *manner* of the action whereby it is produced, or the *cause* which produces it, these can not be so much as aimed at.'

In short, the Laws of Nature simply describe the succession of phenomena, and are reduced to general rules, by means of which we may make very probable conjectures about our further experiences; they are not causal laws. As he argues in Section lvx, 'The *fire* which I see is not the cause of the pain I suffer upon my approaching it, but the mark that forewarns me of it.' Just as for Patricius, so for Berkeley there is no distinction between things as they are and things as they seem, and, for both, the proper course of science is the Babylonian task of formulating the rules of succession of common experiences. But Berkeley does not hold that talk of other things is meaningless or false, rather he argues that the terms appearing in such talk must be understood really as referring to the things of common sense.

The theory of Brodie

A very similar theory of meaning, and of science, was developed in an attempt at the reform of chemistry made in the nineteenth century. Throughout the period chemistry had been bedevilled by doubts about the atomic conception of nature, but at the same time the atomic conception had proved to be a very good guide to experiment, and had given some sort of understanding of phenomena. Sir Benjamin Brodie, professor of chemistry at Oxford, was not only a chemist of distinction but also a mathe-matician of the first rank, and one of the few in the mid-nine-teenth century to appreciate the significance of the mathematical discoveries of George Boole. Brodie decided to try to develop a mathematically organized, phenomenalist chemistry, that would be as fruitful as the atomic theory but without its implausible and indecisive character. What, he asked, were the characteristic chemical phenomena? There are *operations* by which the substances are prepared, and the relative weights they have produced. Chemical reactions are really weight distribution changes brought about by chemical operations. Considered strictly from the point of view of the phenomena, a chemical element is a simple weight, that is a kind of space-occupant which always, throughout all chemical operations, preserves the same weight.

Away with all this atomic claptrap then, the laws of chemistry are nothing but rules for calculating what weight relations will be observed after certain chemical operations have been carried out. Hydrogen is not, on this view, a characteristic kind of atom, aggregated in pairs, in molecules, millions of which make up a sample of the gas. It is hydrogenated space, which always has the same simple weight, the least known simple weight. The method of preparing hydrogen is not to be understood as the way to separate atoms of hydrogen from other atoms, but the way to hydrogenate space. This theory had a certain vogue and provoked a good deal of discussion, but suffered the fate so common to phenomenalistic theories: it was forgotten in favour of the atomic hypothesis. (For details of Brodie's Calculus and its method of operational definition, see Knight, Brock, and Dallas, in *The Atomic Debates*, edited by W. H. Brock.)

All three, Patricius, Berkeley, and Brodie, share the view that scientific knowledge is about what can be perceived. Any alleged thing or process that cannot be perceived cannot be supposed to exist for science. The meaning of expressions which seem to refer to what cannot exist must be treated as really referring to phenomena. And the idea that such terms as 'chemical atom' refer to things we either do not or cannot perceive is just a mistake about their epistemological status. The laws of nature and scientific theories are then nothing but records of past experiences which can be used to anticipate future experiences. The second group of phenomenalist theories shares these opinions in part but differs in not accepting the principle that the ultimate phenomena are what we ordinarily perceive. Again I shall look at three famous versions of this theory.

The phenomena are analysed out of what we ordinarily perceive

Mach's Theory of Elements

Mach's opinion as to the nature of the Laws of Nature does not differ from that of those phenomenalists I have already described. The function of Laws of Nature is the 'mnemonic reproduction of facts in thought', that is, they are summaries of what we have already experienced. Because they are general they point forward to other similar experiences that we might come to have. Since they are summative in character their virtue must lie in their degree of summariness, i.e. in their simplicity and economy. A good theory sums up the facts economically, a bad theory extravagantly. Mach himself devoted considerable effort to the reduction of the number of independent concepts required for mechanics. In his *Science of Mechanics*[1] he shows with some measure of success that both the concept of mass and the concept of mechanical force are not used descriptively of experiences but are expressible in terms of accelerations. Force, as we all know, is defined quantitatively in terms of the product of mass and the acceleration which the force is supposed to produce. Mass can be somewhat similarly treated, since the ratios of a pair of masses are in the inverse ratios of the accelerations they acquire under similar conditions. Mach tries to show that the similarity of conditions can be specified quite objectively without bringing in the idea of similarity of force. Such terms as 'mass' and 'force', which seem to refer to qualities and agencies, which are in themselves not perceivable, can, Mach believed, be paraphrased away. Their function, like the laws in which they appear, is summative and simplifying. What then is left?

The analysis of experience, Mach was convinced, led one to the idea that its ultimate components were sensory elements. An example will show what this means. Where Berkeley would have been content to have recognized the existence of an apple, Mach would proceed further. He wanted to treat the apple as a group of sensory elements, round shape, red and green colour in various patterns, firmness to the touch, sharply sweet to the taste, and so on. 'Apple' is a word whose function in the language is neatly and economically to express in summary form the presence of this

[1] E. Mach, *The Science of Mechanics*, trans. T. J. McCormack (La Salle, Ill.: Open Court, 1960), ch. 2.

and similar groups of sensations, that is of sensory elements. But does not this make apples just groups of human sensations, which is absurd? They are more than that in Mach's theory, and he tried to accommodate this point with an ingenious theory. The elements, he argued, appear under different categories as they are differently related. This just means that when an element, say red colour, is considered with respect to a perceiver, a human being say, who is aware of the colour, it appears as a colour sensation. Shape, weight, smell, and so on, when considered in relation to a perceiver *are* the sensations of various kinds. But when the elements are considered in relation to each other, they are what we call qualities, and together constitute things. The word 'thing' is just an economical way of referring to a particular kind of grouping of elements, considered either with respect to each other, in which case they are the properties and qualities of the thing, or with respect to the person who perceives the thing, when they are the sensations that the perceiver has when he perceives the thing. The experiences which the laws of nature summarize and enable us to anticipate are then sensory elements, sensations, such as feelings of pressure, of roughness, smells, colours in the visual field, and so on. On this theory, we ought not to treat our experience as perceivers of things, nor should we suppose that we have knowledge of things. Knowledge is only of elements, and considered in relation to ourselves, this is knowledge of the order and sequence of our sensations. Sensations are the ultimate phenomena and knowledge of sensations the only true scientific knowledge.

To all this one might well object that this is not what scientists seem to be doing. They do not generally regard themselves as recording their sensations and then looking for patterns in the records which can serve as a neat summary of what has been sensed. Characteristically, they proceed by building apparatus and operating with it. Many experimental operations are measuring operations and their result is numerical. In general the result of a measuring operation is a product of a number and a standard. Measuring operations performed on the floor can give results like 5×1 metre. The idea that we might make measuring operations themselves the ultimate scientific phenomena is the basis of the theory of operationism.

Bridgman's Theory of Operations

The origin of this theory lay in the profound shock that many physicists felt when they realized that they had been using a system of concepts and a comprehensive theory, in the shape of the Newtonian Principles, when on really profound analysis that theory could be shown to be defective and limited. How, then, could a science be devised which would be built up from concepts which would be immune from vast revolutionary changes like that which had overwhelmed physics and astronomy with the advent of relativity? It seemed to Bridgman that the only way in which this could be achieved would be by confining science to the application of those concepts which could be understood in terms of experimental operations, and then, whatever happened to theory, the content of true scientific knowledge would remain unchanged.

All this could be ensured, Bridgman believed, by the device of defining the meaning of all concepts employed in the sciences only in terms of the operations by which measurements were made.[1] 'Length' then ceased to be an attribute of bodies, and was defined as the set of operations of sliding rulers, and marking co-incidences, and counting these operations. The description of the operations was to count as the complete and final definition of a term. Should two different sets of operations be possible to measure a quantity, then in fact, according to this theory, there were thus defined two distinct concepts which had previously been confusedly taken together. As to what a theory was, this problem also fell to an operational analysis, since a theory could be nothing but a means of working out the results of a new set of operations on the basis of some other set. The operations of calculation, 'pencil and paper' operations, became the total empirical meaning of the theory. Instead of conceiving of science as describing things and their properties, powers and interactions, it must be treated as describing the operations we perform in the laboratory and in the study. These are all that science is about, and they become the ultimate phenomena.

This theory was, as we have seen, anticipated in spirit and indeed in many details by Brodie in his attempt at the reformation of chemistry. We shall find that operationism has many un-

[1] P. W. Bridgman, *The Logic of Modern Physics* (New York: Macmillan, 1954), ch. 1.

satisfactory features, some of which you have no doubt thought of in the course of reading this exposition of its main principles. At this point I introduce it as an important example of a particular point of view of the nature of scientific knowledge. Once again, if all that scientific terms mean is to be given in terms of a description of a set or sequence of operations, then all that scientific knowledge can encompass must be no more than knowledge of the sequence of operations. It does not, for instance, include knowledge as to the reasons why the operations are the way they are. Theory means no more than pencil and paper operations by which the numerical values of the results of sequences of operations can be computed.

The operationist viewpoint has also had a certain influence in the science of psychology. In their endeavour to get away from subjective and fallible judgements, psychologists welcomed operationism as a test for the scientific viability of empirical concepts. If no operations could be described by which the presence, absence, and qualitative or quantitative variation of a quality or property or state could be determined objectively, then it was thought that that quality, property, or state had better not figure in a scientific psychology. Thus, by using the test of operational definition, it was hoped that all concepts employed in the science would automatically achieve objectivity. In this, as in other sciences, the effect of operationism was to trivialize theory, and ultimately to trivialize the science itself.

It has often been pointed out that there is a considerable difference in the worth of different experiments. Some experiments advance knowledge, others do not, however accurate they may be and however carefully performed. The usual way of deciding whether an experiment is worth while involves reference to a theory. But if a theory is nothing but a calculatory ancillary to the operations of the laboratory, it can give no guide as to the relative value of the experiments. In this situation all experiments have equal point or, an unkind critic might maintain, equal lack of point. In this way, any describable operations could come to seem to have some scientific value, since, being operations, their description defines, according to Bridgman, a descriptive concept or concepts. And yet it is quite clear that many operations produce entirely trivial and often pointless results. Indeed, the *reductio ad absurdum* of the theory is not hard to find. Any random hooking-up of any random collection of equipment can be described, and doing anything whatever with it can be described, in this permissive atmosphere, as 'operating' with it. In this way, an infinite number of empirical, 'scientific' concepts can be trivially generated, one for every distinct random assemblage of equipment, whether or not it actually yields any results, since there is not even a requirement in the operationist theory that any experimental set-up should actually work.

Eddington and the Ultimate Observer

An even stranger conception of the ultimate phenomena is to be found here and there in the philosophical works of the great astronomer and mathematician, Eddington.[1] His ideas seem to derive from two main principles. The first is that the best data are numerical, the second that the actual task of the observer, at least in physics, is to notice the coincidence of the pointer with a graduation on a scale. It is a short step from there to the view that the numerical readings of scales are the fundamental phenomena. In this brief passage of thought, as I have presented it to you, you will have already noticed a slide very characteristic of phenomenalism.

It seems not unreasonable to argue that the basic data are the numerical readings of instruments with scales. But the conclusion of the passage of thought is that the fundamental phenomena are the numerical readings of scales. We pass from n being a measure of some property or power or state P, to the view that n itself is the phenomenon, and P is eliminated from the system of concepts. But without P we are unable to express the point of obtaining the measurement n. Eddington's detailed argument begins with the development of the idea that 'the initial data of physics are nerve signals'. The human brain decodes the signals and builds up a picture, or forms a conception of the causes of these patterns of signals. The whole process is perception. Eddington then seems to take the analogy between nerve impulses and brains on the one hand, and stimuli upon instruments, and instruments being read by physicists on the other hand, as practically an identity. The instrument readings provide the data, like nerve impulses, and the

[1] A. S. Eddington, *New Pathways in Science* (Cambridge: Cambridge University Press, 1938), ch. 1.

physicist interprets them and so constructs a world picture, like a person perceiving.

This muddled analogy is not confined to the thought of Eddington. It is often taken as a commonplace by physicists, and appears quite frequently in discussions of the philosophy of quantum mechanics and special relativity. It sometimes appears in such remarks as assert that modern physics differs from the old physics by including the observer in the system. This, of course, is quite untrue. Sometimes the basis for this mistaken belief is the analogy which misleads Eddington, sometimes it is a confused appreciation of the fact that many observations and many acts of measurement interfere, sometimes irreversibly and sometimes unpredictably, with the system being studied. I want to consider now only the alleged analogy between nerve impulses and brains, and readings and observers.

After a moment's scrutiny the analogy dissolves, since the brain is not aware of nerve impulses in the way in which observers are aware of their data. People are aware of things and their awareness is not in the least mediated by any process of understanding nerve impulses. Our awareness of things and processes is quite unaffected by our knowledge or ignorance of the physico-chemical mechanisms of perception. There simply is no analogy at all. A person is what is aware, not a person's brain or any other organ, and the data of his experience are what he perceives, and not any part of the mechanism by which he perceives. The physicist reading his instrument is as much a person perceiving as you are when you look up and see the scene outside the window. But *unlike* the person perceiving, physicists do have to make inferences from the reactions of their instruments as to what is affecting them and causing them to behave in the way they do. By contrast, people perceiving do not infer from nerve impulses, not even from bare sensations, what is causing them. They perceive the things in the world without sensory or nervous intermediaries. Indeed one might turn the tables on Eddington and point to many cases in which the immediacy of the reaction of an instrument is so striking as to bring a perceptionlike treatment of instruments into physics. And instead of speaking of detecting a magnetic field, or the path of an electron, we sometimes speak as if the pattern of iron filings were the field made visible, and as if the cloud-chamber track of droplets, condensed upon ions, were the very track of the electron.

And, in a sense, these perceptual treatments of detecting devices do have a certain merit. What has no merit at all is the transfer of a false inferential picture of perception into the epistemological understanding of the workings of instruments.

Phenomenalism just will not do, and all its disguises and forms increase our dissatisfaction. There are no data which are recognizably basic, and nothing is gained by the drastic truncation of science which would construe all terms, including those which figure in explanatory theory, as constructions out of terms for basic data, or phenomena, and as meaning nothing but that. 'Atom' does not mean 'proportion of substance reacting', nor does 'gene' mean an invariance relation between certain statistical features of the distribution of characters in successive generations of adult organisms. So we must look askance at the model of mechanical force for understanding the structure of all scientific theories, and turn rather to models like 'bacterium', 'gene', 'chemical atom'. The capillaries were seen by Harvey to be a necessity if his theory was to be viable, but not *seen* until Malpighi's experiments after Harvey's death. Theory is always *extending* our conception of phenomena, and of what there is in the world.

Hypotheses as fictions

Instead of trying to make all kinds of knowledge into one kind of knowledge, with a quite implausible ideal of reduction, most philosophers and scientists are content to accept the fact that there are at least two main kinds of scientific knowledge, our knowledge of fact and our theories. As we shall become increasingly aware, this distinction is not a dichotomy, that is no hard-and-fast line can be drawn distinguishing fact from theory; but at least we recognize the difference between 'A blue precipitate appeared at 28° C.' and 'Atoms with six or seven electrons in the outer shell tend to form covalent bonded molecules,' even though theory present and the ghosts of theory-past infect the former, and facts are not irrelevant to the latter. But under what picture is this difference to be understood? In this section I am going to consider the idea that theories are fictions, that is that they are to facts roughly as novels are to histories. They read like factual accounts, and ostensibly work according to the logic of description and reference, but they are works of the imagination and nothing else.

A novel depends for its plausibility on using an apparatus of historical reporting which in real life is under the control of truth, but in the novel is not. The episodes of a novel are coherent, and they may be quite convincing, i.e. have the ring of truth, but the key point of difference is this: the people who figure in the novel are not real people: their names and addresses do not refer to real people and places. The art of novel writing consists in constructing characters and endowing them with characteristics so plausibly that we are willing to treat them as real people, that is to suspend disbelief. And we do this partly in the interests of entertainment, and partly for the insight we get into character, because the laws of nature, as it were, according to which the characters of fiction are delineated, must be truths about real human characters.

The fictionalist view of theories takes over all this to science. As an aid to thinking, a plausible theory may be a very powerful tool, and by using known laws of nature in describing the behaviour and nature of the entities with which it deals it gains plausibility. But the entities themselves have no more reality than the characters of fiction, and the terms which are used to describe and particularly to refer to them are like the names and addresses of characters in novels. This idea appears quite frequently in the history of the philosophy of science, and is particularly common when there is some kind of crisis in a science where the current theories are many, and there seems to be no way of resolving the issue between them. Then there seems to be a strong tendency for many philosophers and scientists of that period of crisis to try to resolve the difficulties to which the situation gives rise by classifying all the competitors as fictions, and to attempt no more than a pragmatic or heuristic distinction between them. Ideal theoretical knowledge then becomes the best set of fictions, and these are understood as being the neatest, the shortest, and the most elegant. The search for truth, for the confirmation of one theory at the expense of others, is abandoned. Truth is claimed to reside only in the relations of phenomena. There are thus held to be two radically different aims and ideals, and so radically different criteria of excellence.

Three important episodes of this sort have occurred in the history of science so far. The one I shall deal with here was the conflict in astronomy in the sixteenth century, when the rival theories of Eudoxus, Ptolemy, Tycho Brahe, Copernicus, and Kepler were all known, and each of them had passionate adherents. One way out of the impasse created by the fact that their conflict seemed irresolvable by any further observations, was the view that they were not competing as accounts of the true structure of the heavens, but competing fictions whose advantages were to be judged solely by their utility and their beauty.

Another important episode of a like kind centred around the problem of the status of atomic theories in chemistry, and involved a debate which continued from about 1810 until the beginning of the twentieth century. Many chemists held that the atomic theory in chemistry was a nest of useful fictions, but was not to be considered as either true or false. The latest episode of this kind has occurred in our time in fundamental physics. It seemed to become more and more difficult to find a single adequate picture of the nature of the entities which were coming to be thought of as the basic constituents of matter. In certain circumstances they seemed to be particlelike, in others wavelike. A fictionalist view grew up when the attempt to reconcile these two pictures seemed less and less successful. It was called 'complementarity'. Electrons which in some circumstances appeared particulate, could be diffracted when in sufficient number in a beam. Light, which since the eighteenth century had been generally agreed to be a wave motion, had for certain purposes to be supposed particulate. The wave picture and the particle picture were then supposed to be complementary, and each fictional. Consistency was preserved since the two incompatible pictures need never be applied together. To some extent this fictionalist attitude to the problem of finding out the nature of the sub-atomic entities has persisted to the present day. No doubt these views will seem as obtuse to scientists of the twenty-first century as do those of the rival astronomers of the sixteenth century to us now.

A very typical exposition of the fictionalist point of view is to be found in the *Fundamentum Astronomium* of Nicholas Raimarus Ursus, published in 1588. Heading the chapter with a quotation from Hosea, 13, 'I will go against them like a bear', a reference to his own name Ursus, he offers an account of the role of fictions in astronomical theory as follows:

A hypothesis or fictitious assumption is an imaginary picture of certain imaginary circles in an imaginary model of the system of the universe, capable of accommodating observations of the heavenly motions, and invented, assumed and introduced for preserving and saving the motions of heavenly bodies and for expressing them in quantitative terms. I say an imaginary picture of an imaginary model of the system of the Universe, not a true and genuine one for that we cannot know. It is a picture, not of the system itself, but of such a model of it as may be imagined by the mind and embraced conceptually. The hypotheses we invent are nothing more than fabrications which we imagine and construct concerning the system of the Universe. Therefore it is quite unnecessary, nor should it be demanded from the devisers of hypotheses, that those hypotheses should correspond altogether, and in all repects, and in every way to the system of the Universe itself (for such hypotheses cannot, I think, be constructed), and you may take it that everywhere, and in every form of constructed hypothesis of which a great many such forms can be imagined and presented, certain very stupid absurdities remain, provided only that they agree with and correspond to the quantitative aspects of the motions and not the motions themselves, and provided that the quantitative relations of the heavenly motions can be preserved and saved by them. For otherwise they would not be hypotheses or (what is the same thing) fictitious assumptions: but they would be true (and not invented) representations of the true (and not imaginary) form of the system of the universe. And so hypotheses cannot be faulted for being assumed contrary to the common principles of other arts and sciences even if they be assumed contrary to the infallible and most certain authority of the Holy Scripture. For it is the function of hypotheses to investigate, to hunt down, and to elicit the true answer to the question, by means of fictitious or false assumptions. For it is permitted and allowed astronomers as a sort of astronomical licence to invent hypotheses, whether they be true or false and fictitious, of such a kind that they may be sufficient for the phenomena and appearances of the heavenly motions, and may duly exhibit their quantitative relations, and in this way may achieve the goal and target of their art. In just the same way as tends to be done in most other branches of learning, in which, for the most part, not truths, and not even probabilities, let me tell you, are assumed, but those assumptions are wisely made which yield the most useful results. (Translation by T. I. M. Beardsworth.)

With this was associated an even more extreme view. One might, if pressed as to the truth or falsity of a novel, have to admit that in a certain sense it is false. There are some sophisticated, logical theorists who would create a limbo of 'neither true nor false' into which such fictions fit and while this may be, in the long run, the best way to handle discourses to which there corresponds no reality, many people would really wish to say that fictions are false. This idea has been imported into the discussion of the philosophy of science. The logical model for a theory then becomes something like this:

1. All birds are mammals
2. All mammals are feathered
3. All birds are feathered

1 and 2 make up a theory which explains the fact that all birds are feathered, and it is a good theory because it implies a true conclusion. Both premisses are fictions, because both premisses are false.

The criterion for a theory becomes logical, not empirical. And if we can devise a theory meeting the logical criterion of validly entailing the known facts and laws, and validly entailing consequences which are shown to be true by observation, and which is also simple, elegant, economical, and beautiful, so much the better. And since in astronomy there are five theories, all of which entail the known facts (that is from known initial conditions the past and future positions of the heavenly bodies can be inferred according to the theory), and all of which satisfy the logical criterion, the only choice that can be made between them must rest upon their simplicity and beauty. We are missing the point, on this view, if we ask which theory is true and which false, because the aim of theorizing is to create the most satisfying fictions.

This view was strongly disputed in the sixteenth century, and as spokesman for the critics I choose Christopher Clavius. Here is what he has to say against the fictionalists in his book *In Sphaeram Ionnis de Sacro Bosco* (Lyon, 1602), pp. 518–20:[1]

Finally we may conclude our topic as follows: just as in natural philosophy we arrive at knowledge of causes via their effects, so too in astronomy, which has to do with heavenly bodies very far distant from us, we can only attain to knowledge of the bodies themselves, of how they are arranged and constituted, through study of their effects, that is their movements which are visible to us. For just as natural philosophers have inferred, with Aristotle, from the alternating birth and

[1] This popular work went through eleven editions between 1570 and 1618.

decay that occurs in nature, the existence of a prime matter together with two principles of natural change, and many other things, so too astronomers through studying the varying motions of the heavens from sunrise to sunset, and from sunset to sunrise, found a fixed number of heavenly spheres; some have said eight, because they found only eight different kinds of motion, others ten, having noted ten different kinds; in the same way, using other observations, they have determined the arrangement of the heavenly spheres, as we expounded at length in Chapter 1. It is therefore convenient and entirely reasonable that astronomers should look to the particular motions of the planets and to the various appearances, to inform them of the number of particular spheres that carry the planets round with such varying motions, and their arrangements and shapes: on condition however, that causes can thereby adequately be assigned to all the motions and appearances and nothing absurd or inconsistent with natural philosophy can be inferred therefrom. Wherefore, since eccentric spheres and epicycles enable the astronomers effortlessly to preserve all the appearances, as is clear partly from what has been said, and may even more plainly be under-stood from their theories, and since no absurdity follows from them, and nothing inconsistent with natural philosophy, as will soon be agreed, once we have dealt with the objections that tend to be levelled against such spheres by their opponents; justifiably therefore have the astronomers concluded that the planets travel in eccentric orbits, and epicycles, not in concentrics since the latter do not enable us to preserve all the manifold variety displayed by the planets in their motions.

But our opponents try to undermine this argument, saying that while they concede that all the appearances can be saved by postulating eccentric spheres and epicycles, it does not follow from this that such spheres are to be found in nature; on the contrary they are wholly fictitious. For (1) perhaps all the appearances can be saved in a more convenient way though we have not yet discovered it; moreover (2), they may be truly saved by the above mentioned spheres without the spheres themselves being any the less fictitious, or in any way the true cause of those appearances, just as one may reach a true conclusion even from a false premise, as is evident from Aristotle's 'Dialectic'.

We can add weight to these objections from the following considera-tions: Nicholas Copernicus, in his book *De Revolutionibus*, preserves all the appearances in another way, by postulating, of course, that the firmament is stationary and fixed, and the Sun too is fixed in the centre of the Universe, and by assigning three-fold motion to the Earth in the third heaven. So eccentrics and epicycles are not necessary for preserv-ing the appearances in the case of the planets. Again, Ptolemy, by means of the epicycles, provides a cause of all the appearances in the case of the Sun, and preserves them by means of the eccentric. So it cannot be concluded from our third argument that the Sun moves in an eccentric, since perhaps it travels in an epicycle.

Nonetheless it must be said that our third argument retains its force, and the objection of our opponents is inconclusive. For firstly if they have a more convenient way of saving the appearances let them show it to us, and we shall be satisfied and thank them heartily. For the Astronomer's one concern is that all the appearances in the heavens should be saved in the most convenient way, whether this be done by eccentric spheres, and epicycles, or in some other way. And because no more convenient way has been found up to now than that which preserves everything by eccentrics and epicycles, it is certainly credible that the heavenly spheres are composed of circles of this kind. But if they cannot show us a more convenient way then they should at least accept this one, based as it is upon such a variety of appearances : unless they wish not only utterly to destroy natural philosophy as it is expounded in the academies, but also to prevent access to all the other arts which study effects in order to discover causes. For whenever anyone infers some cause from its visible effects, I will say just what my opponents do : 'Doubtless another cause, at present unknown to us, can be found for those effects.' Or if one should accept that what has been found is the actual cause, because it has some connection with the effects from which it has been inferred, then eccentrics and epicycles must also be allowed as causes, seeing that they have so close a connection with the appearances that they are able to preserve them all easily by means of their motions. Next if it is not right to conclude from the appearances that eccentrics and epicycles exist in the heavens, because a true conclusion can be drawn from false premisses, then the whole of natural philosophy is doomed. For in the same way, whenever someone draws a conclusion from an observed effect, I shall say 'That is not really its cause. It is not true because a true conclusion can be drawn from a false premise.' And so all the natural principles discovered by philosophers will be destroyed. Since this is absurd, it is wrong to suppose the force and weight of our argument is weakened by our opponents. It can also be said that the rule of dialectic, *that truth follows from falsehood*, is irrelevant, because it is one thing to infer a truth from a false premise, and quite another to preserve the appearances by means of eccentrics and epicycles. For in the former case it is by virtue of the syllogistic form that a true conclusion is drawn from a false premise. (Translation by T. I. M. Beardsworth.)

Scepticism about hypotheses

The main motive for treating hypotheses as fictions has always been the alleged impossibility of telling which hypotheses are true and which false. Many, indeed infinitely many, different sets of hypotheses can be found from which statements describing the known facts can be deduced, so any criterion based upon the logical power of hypotheses is bound to be equivocal. But instead of going so far as to declare that all hypotheses are fictions, some philosophers and scientists have taken the line that even though it does make sense to consider that one of a set of rival hypotheses is true, it does not make sense to pursue the question as to which one. The choice among hypotheses would be made as the fictionalist advocates, by reference to their logical power, mutual coherence into theories, simplicity and economy, elegance and beauty, and so on. The hypothesis which was best according to these criteria might, by chance, be true, but it equally well might not. The referring expressions, nouns, names, etc., which in an ordinary discourse would refer to existing entities, cannot be distinguished from fictions, which only pretend to refer. We can no more settle the question as to the existence of things, qualities, and processes referred to in a hypothesis, than we can settle the question of its truth. To settle the truth of a hypothesis we have to know whether the things it refers to exist, because only if they do can they be studied to see if they have the qualities, natures, and behaviour that the hypothesis alleges they do.

In the great astronomical debate of the sixteenth century the sceptical position was not uncommon, and it had its most famous exposition in the preface to Copernicus' book *De Revolutionibus*. Kepler, who was passionately opposed to scepticism, discovered that the preface was not written by Copernicus, but by one Osiander, who was responsible for the final stages of the printing of the *De Revolutionibus*. Copernicus himself was a realist, as can be seen from the views attributed to him, no doubt accurately, by his pupil Rheticus in his book *Narratio Prima*.[1] Here is a précis of the unsolicited preface Osiander contributed. Kepler found a letter of Osiander in which he avows the intention of drawing off the fire of conservatively minded persons who would have been shocked by an advocacy of a heliocentric theory as true. The précis was made by Ursus, and appears in his *Fundamentum Astronomium*.

It is the function of the astronomer to compare the history of the heavenly motions by diligent and careful observation: then to devise

[1] For a translation of that interesting book, see E. Rosen *Three Copernican Treatises* (New York: Columbia University Press, 1939).

and invent causes or hypotheses of any kind he likes since it is impossible for him by any means to reach true ones. By the assumption of these hypotheses, those same motions can be correctly calculated in accordance with the principles of geometry, both for the future and for the past. For it is not necessary that the hypotheses be true, indeed they need not even be probable : but this one thing is sufficient, that they should provide a calculus consistent with the observations, for it is clear enough that this art is utterly and completely ignorant of the causes of the unequal motions of appearances. Whatever hypotheses astronomy devises (and it certainly devises *AS MANY AS POSSIBLE*), it certainly does not invent them in order to persuade anyone that they are true; but merely that it might correctly yield the numerical relations. Now since various hypotheses are offered from time to time concerning one and the same motion (as with the motion of the Sun, the hypothesis of eccentricity and of the epicycles) the astronomer will prefer the one which is as easy as possible to understand. The philosopher perhaps will rather demand probability. Neither of them however will understand or propose anything of certainty, unless it is revealed to him by Divine inspiration. And nobody, as far as hypotheses are concerned, should expect any certainty from astronomy, since the art itself is incapable of furnishing any such thing, lest by treating fictions as truths and appropriating them for some other purpose he depart from this science more foolish than he came to it. (Translation by T. I. M. Beardsworth.)

A somewhat similar position has been advocated in recent years by certain philosophers and has come to be called 'instrumentalism'. It advocates the view that theories are not to come up for judgement as true or false, indeed they cannot so come up, but are to be judged by whether they are successful or unsuccessful 'instruments' for research. It is interesting to note that this idea, indeed the very name, is to be found in Gassendi's theory of science, set out in his *Syntagma* of 1658,[1] where he declares that all hypothetical or conditional statements should be considered to be 'natural instruments', i.e. devices by which knowledge can be made more orderly and penetrating.

As to the meaning of the terms in theories, the sceptical view can differ little from the fictionalist view. On neither idea of science do the theoretical terms refer to anything existing independently of the phenomena the theory explains. But even if the sceptic does not wish to deny that some terms may in fact refer to

[1] P. Gassendi, *Syntagma* (Lyon, 1658).

real entities and processes, he wishes to insist that we cannot know which, if any theoretical terms do so refer, and so whether a term refers to something real or not cannot be something which in any way enters into the meaning of the term. Scepticism, so far as I can see, must share with fictionalism the idea that the terms that appear in theories get their meaning from their role in a story. The term 'chemical atom' would acquire its meaning in much the same way as does the name Mr. Pickwick. Chemical atoms are things of a sort, in a certain fictional role, as Mr. Pickwick is a man of a sort in a certain fictional role. A fictionalist can be thought of as taking *The Pickwick Papers* for his model of meaning, and the sceptic perhaps can be understood as taking Homer for his. Ulysses might have existed, but our understanding of him derives from the Homeric tale, not from acquaintance.

The examples and the Philosophical Theories compared

If we now think back on the examples described at the beginning of this chapter it is clear that some of them seem to fit in with some one of these three non-realist views of hypotheses, while others do not. The theory of mechanical force seems the example best suited to provide a model for theory construction which looks plausible under some of these heads, and particularly under 'complete phenomenalism'. It does look as if 'force' is nothing but a way of speaking about mass acceleration. Again the light ray seems, at first sight, to fit 'hypotheses as fictions', since it is pretty clearly a fiction, and one has to penetrate the physics of optics very deeply to come to see it as anything else. Finally the 'heat' of calorimetry seems to provide a plausible model for the sceptical point of view. 'Heat' could be treated like 'force', as just a way of referring to the produce of mass, specific heat, and temperature change, but it could be treated like the light ray as a fictional entity, a kind of weightless fluid, which allows us to make sense of the experiments. Or its existence could be treated like the position of the sun in the Copernican system, as something which could be true, could be as described, but about which there is no way of deciding, so that the meaning of the word 'heat' is arrived at by considering its role in a story.

Realism

The choice of another kind of example leads us to another view of theories. Neither the phenomenalist, fictionalist, nor sceptical account seems at all plausible for the story of the virus. There is a fourth view of theories under which the virus theory does make sense. This is the view that the statements of the theory are true or false, and that many of the entities referred to in a theory do exist. They are as much in the real world as are human beings, houses, stones, stars, and so on. This view I shall call 'realism'. Now a realist does not maintain that every hypothetical entity exists, nor does he maintain that every statement in a theory is to come up for instant judgement as either true or false. All he needs to maintain is that some hypothetical entities are real, and that since some theoretical statements have been judged true or false, other such statements can, in principle, come under that sort of judgement. To make the realist position clear two important concepts need to be introduced. These are *reference* and *demonstration*. We can get some idea of the notion of making reference from the business of using proper names. A proper name like 'Henry' can be used to make reference to someone whether or not he is present. Indeed part of the value of having proper names in a language is just this, that they can be used in conversation between third parties to refer to someone who is not present during the conversation. The same is true for other kinds of words. Nouns, for instance, can appear in phrases whose function is to enable us to refer to things, whether or not they are present. Referring to entities is, then, something we do with words. Since words can be understood in isolation from the entities that they are used to talk about, verbal reference can be made to things whether or not the things can be observed or presently shown to us, provided we can understand the meanings of the words used to make reference.

'Demonstration', and its close relative 'indication', are typically acts performed with or by a gesture, in which an entity is pointed out. An entity cannot be indicated in its absence, only in its presence. A typical indicating gesture is that of pointing. To be able to indicate something is the final, incontrovertible proof of its existence. It should be clear that we can refer to many things we cannot indicate, but that if we can demonstrate a thing to which we have previously made reference, on this or other occasions, then we have proved that thing exists.

In a theory there are theoretical terms, and they can be used to make verbal reference to hypothetical entities, whether or not these entities can be observed or be present to our experience. Not all theoretical terms are what they seem. Some *are* really nothing but picturesque or shorthand forms of complex expressions whose meanings are different from that which at first hearing one might suppose them to have. The best example of this is 'force'. But if, in addition to verbal reference, an act of demonstration can be performed by which a hypothetical entity is pointed out or indicated, then that entity must be said to exist. The realist position can now be set out schematically, and consists of the following principles:

1. Some theoretical terms can be used to make reference (verbal) to hypothetical entities.

2. Some hypothetical entities are candidates for existence (i.e. some could be real things, qualities, and processes in the world).

3. Some candidates for existence, for reality, are demonstrable, i.e. can be indicated by some sort of gesture of pointing in the appropriate conditions.

To illustrate these points let us consider the logical history of some theories of disease. Since the seventeenth century the idea that disease is due to infection has steadily gained ground. In origin there were two such theories, that of Van Helmont, the theory of the alien *arche*, and the theory of noxious vapours. In Van Helmont's theory a disease was caused by an alien *arche*, or physical principle of operation, taking over the control of the host's organs and body, and forcing them to operate in its way. The alien *arche* was inhaled or ingested in the form of micro-organisms. In the noxious vapour theory the same agent that was responsible for bad smells was, when inhaled responsible for disease. Both theories were intended realistically, and both '*arche*' and 'noxious vapour' were terms which purportedly referred to actually existing entities. In the nineteenth century it was shown that the presence of the noxious vapour was not a necessary condition for the disease to occur, nor was it a sufficient condition. In the classical malaria investigations the last stronghold of the theory was taken, by the application of methods of inquiry like Mill's. There was no doubt that there were gases which were responsible for bad smells but they were not the causes of disease.

By this time Van Helmont's theory had developed into the 'bacterium' theory. Bacteria were regarded as very small parasites which invaded the body and upset its metabolism thus causing the symptoms of the disease. Verbal reference to these putative agents was made possible by the introduction of terms like 'bacterium', but their existence was not conclusively established until it was possible to demonstrate them with the microscope. In this way they were shown really to be micro-organisms. In the careful experiments of Pasteur and Koch the presence of micro-organisms was shown to be the cause of many common diseases. But not of all. For those for which no bacterial cause could be found an old word was adapted to express a new concept. The *virus*, conceived after the model of the bacterium, became the putative agent responsible for those diseases for which micro-organic causes could not be found. Verbal reference to these hypothetical entities was possible because of the presence of the word 'virus' in the language. Finally demonstration of them was achieved by the use of the newly invented electron microscope, whose enormous powers of magnification and resolution made this possible. From candidates for reality they passed to the status of real things. Twice, therefore, in the history of theories of disease, has the process which the realist regards as typical occurred.

If mechanics with its eliminable concept 'force' provides a model science for phenomenalists, the virus theory of disease provides a counter-model for realists. Science, it seems, contains both kinds of theory. The resolution of this confrontation is beyond the scope of this book. We shall recognize many kinds of theory, from the wholly phenomenalistic to the wholly realistic.

One must recognize some degree of flux in the status of the concepts in use in theories. A new concept is not introduced and then permanently fixed in a certain logical and epistemological status. A new term may be taken to refer to an entity which could be taken to be a candidate for reality, that is the question of its existence could be taken seriously, and then a shift in the organization of theories in that science could deprive it of that status. 'What!' they all say. 'You did not take the Freudian concept of the unconscious seriously surely? It is nothing but a metaphor for the idea of emotional habit.' This sort of change occurs between the status of fictions and of candidates for reality. And it occurs partly in a shift in the climate of opinion, a shift in the conception of the possible: and partly through changes in the circumambient theories. The shift in the status of Mendel's fiction, the fixed genetic factor, into a candidate for reality, the gene, has occurred in part through the development of the theory of the cell, and indeed through greater knowledge of the processes in cell division, and all this has been supplemented by a mounting array of biochemical theories culminating in discoveries which have allowed the identification of the gene with a certain molecular structure within one of the components of the cell. And I suppose it is changes in the conception of the possible that have altered the status of the elixir of life from a candidate for reality (it was a fluid worth seeking) to a fiction (it is not something that could be found). Oddly enough some of the conceptions from the Ancient World which still seemed possible in medieval times, and which then ceased to seem possible in the Newtonian age, have in fact been realized. The transmutation of the elements, the production of gold from a base metal, is a hypothetical process which has been realized though not as a commercial proposition. Perhaps the development of biochemistry will lead to the realization of the elixir of life. Any given theoretical concept is temporarily fixed in a certain status by a complex of factors which can usefully be discussed only in the particular case.

Under quite different influences the boundary between demonstrable candidates for reality and non-demonstrable candidates can also shift. Once a certain entity becomes a candidate for reality the question next arises as to how it can be identified. Whether or not a certain thing, process or quality can be identified and its existence demonstrated depends on the state of instrument technology. There are two possible lines of development.

First, finer and finer probes can extend the sense of touch, amplifiers the sense of hearing, and more and more powerful microscopes can extend the sense of sight. Some candidates for reality become demonstrable because of the developments of instruments of this kind. The blood capillaries which complete the circulation as postulated by William Harvey were made demonstrable by the development of the optical microscope, as indeed were viruses by the development of the electron microscope.

On the other hand, when a new entity is introduced as a candidate for reality which is not like the sorts of things which can be sensed, it is not for that reason undemonstrable. It may be that an instrument can be devised to *detect* it. The simple electroscope is a device for detecting electric charge, and the divergence of the gold leaves demonstrates the existence of a charge. But a charge is not something we would ever expect to observe by any extension of the senses. At best we can feel the prickling of a small discharge. With magnetic fields even this small sensory bonus is denied us, but the way magnets orient themselves in the field makes this demonstration as certain as you like.

These developments depend both upon the inventiveness of instrument makers and upon the existence of theories according to which suitable instruments can be made. The detection of radio-stars depends not only upon the ingenuity of the inventors of the radio-telescope but also upon the theory of electromagnetic radiation, by means of which the reactions of the instrument are interpreted as due to the radiation from particular sources in the heavens. But often by successfully extending the senses, theory is, as it were, eliminated. The existence of bacteria as demonstrated by the optical microscope is quite unaffected by changes in the theory of the instrument. However, even in the case of detecting instruments all is not wholly dependent on theory. If an instrument is quiescent for many situations and then suddenly twitches into activity (the gold leaves diverge, the litmus paper changes from blue to red), *something* is affecting it. There is an electric charge, the water does contain some acid (hydrogen ions). But what an electric field *is*, what is its 'essence and quiddity' as Bacon put it, just what is an acid; the answers to these questions are subject to change and fluctuation as theories develop and decline and as the general view of what is possible in nature alters, and the ideas as to what are the basic structures of things change.

The poles of epistemology of science, phenomenalism and realism, are distinguished by the sort of knowledge they admit as scientific. Phenomenalists see science as consisting exclusively of general laws of phenomena, of the form 'All events of a certain kind are followed by events of a certain other kind'; as, for instance, 'All impacts of particles are accompanied by changes of velocity according to the momentum conservation law.' Realists see scientific knowledge as consisting fundamentally of knowledge of the existence of things and materials of certain kinds, and knowledge of their constitutions and natures. From this knowledge, laws of the behaviour of things, that is empirical laws, flow. For most of scientific knowledge realists are surely right to see the extension of our knowledge as alternately an extension of our knowledge of what things and materials there are, and an extension of our knowledge of their natures. In this chapter we have come to see the variety of ideals of knowledge, and to understand not only that existential hypotheses are proper but how they are determined as well. We have entered here, in a preliminary way, one of the most profound and difficult parts of the philosophy of science. Here the two great systems of thought, positivism and anti-positivism are brought into relation, and we see how together they provide a spectrum of the forms and content of knowledge.

Summary of the argument

1. Examples of science

(a) The use of the concept of 'chemical atom' to explain the observed behaviour of materials.

(b) The use of the concept of 'light ray' to give geometrical form to optical discoveries. The concept has been given two quite different interpretations, depending upon which general theory of light has been dominant.

(c) The use of the concept of 'heat' in the explanation of simple calorimetry.

(d) The use of the concept of 'mechanical force' in the formulation of the principles of mechanics in Newtonian style.

(e) The use of the concept of 'virus' to explain the appearance of diseases with no apparent bacterial cause.

(f) The several theories to account for the retrogradations of the planets.

(i) Babylonian zigzag functions.

(ii) The concentric spheres theory of Eudoxus and Aristotle.

(iii) The epicyclic theory of Ptolemy.

(iv) The theory of Ursus and Tycho that the planets are a sun-centred system the whole of which is in motion around the earth.

(v) The theory of Copernicus that the planets are in motion around the sun in circular orbits on eccentric centres.

(vi) The theory of Kepler that the planets are in motion around the sun in elliptical orbits.

Note: Each of these theories provided an adequate mathematical basis for the computation of the ephemerides, that is each saved the phenomena, and explained retrogradation. There was no empirical way of deciding between them.

Our programme will be to describe the main epistemological theories and to consider them against what can be supposed to be known in each of the examples of science cited above.

2. Complete phenomenalism

Only propositions about observed phenomena have the status of genuine knowledge. On that theory science would concern itself only with the identification, classification, and codification of phenomena.

(a) The phenomena are the relations and behaviour of ordinary things.

(i) The theory of *Patricius*: There is no sense in trying to maintain a scientific distinction between what really happens and what seems to happen. This alleged distinction is defeated by the principle: 'Only what can be observed exists.' It is interesting to notice that this principle has been revived in modern times in logical positivism through the verification theory of meaning, and is used in defending the Copenhagen Interpretation of Quantum Mechanics.

(ii) The theory of Berkeley: The theory is based upon a number of principles:

(A) There is no distinction to be drawn between real things and a person's perception of things, or the ideas of them in his mind. The abolition of the distinction allows him to claim that all experience is of ideas.

(B) Only spirits can be causes, since causation requires activity, so there is no causal structure to the ideas we experience. Their cause is to be found in God's activity, when they are regular, and in our own when they are not.

(C) Science then must be confined to the identification of regular sequences of ideas.

It follows from these three principles that theoretical concepts which seem to describe causes, such as 'force of attraction', merely redescribe the phenomena.

(iii) The theory of Brodie: The phenomena consist both of the quantitative and qualitative changes that occur in nature, naturally or by manipulation, and the operations required to bring these changes about. The laws of nature must therefore really describe nothing but the sequences of operations for bringing about specific changes in qualities or weights of materials. Theoretical terms like 'atom' are really only devices for redescribing the phenomena succinctly, and have no other meaning.

(b) The phenomena are analysed out of what is ordinarily perceived.

(i) The theory of Mach: Mach shares the views of the above phenomenalists as to the function of the laws of nature, namely economic description of the phenomena, reducing as far as possible the number of independent concepts required. The phenomena, though, are sensory or qualitative 'elements'. These elements are sensations when considered with respect to a person, and qualities when considered with respect to each other. 'Thing' concepts serve only to express regularities of coexistence and compresence of 'elements'.

(ii) The theory of Bridgman: Like Brodie, Bridgman wished to include manipulative operations in the basis of science. Unlike Brodie he did not conceive them as changing this or that quality, but as yielding a set of numbers. The quality or property of which the yielded set of numbers was the measure was defined by Bridgman by the set of operations. Theoretical concepts are eliminated in favour of 'pencil and paper operations' upon the sets of numbers yielded by apparatus.

Examination of the theory of Bridgman: The fact that certain sets of operations yield similar sets of numbers is a mysterious coincidence, and on a strict interpretation of the theory cannot be explained by the hypothesis that they are each measures of the same property of a thing.

It further follows that any random hook-up of equipment and sequence of operations defines an empirical concept.

(iii) The theory of Eddington: The ultimate facts are sequences of numbers produced by simple observations of pointer coincidences. On the basis of this the observer forms a mental picture of the world. The theory is based upon the alleged analogy between nerve impulses and their transformation into perception by the brain, and the observation of pointer coincidences and their transformation into data by the physicist.

Examination of the theory of Eddington: The analogy fails because the brain does not make inferences as to the world from the nerve impulses nor does a person infer a world from his sensations. We need a perception theory of instruments, rather than an instrumental theory of perception.

3. Hypotheses as fictions

Theoretical statements are not reducible to statements descriptive of observations, but their role is only to give order and system to statements about observations, and the concepts employed in them are similar to those employed in works of fiction.

(*a*) Concepts used exclusively in theory share the grammar of the concepts used in describing observations but do not refer to real things and processes.

(*b*) Fictions must have some degree of plausability, which they gain by being constructed in the likeness of real things.

(*c*) It is inappropriate to demand that theoretical hypotheses be true.

(*d*) There is no genuine conflict between several different theories to explain the same facts, e.g. between the various sixteenth-century astronomical theories, or between the complementary wave and particle conceptions appropriate to different situations in subatomic physics, cf. N. Ursus, *Fundamentum Astronomium*, N. Bohr, etc.

(*e*) Taken strictly all statements in a work of fiction are false, it has been argued. This led in the sixteenth century to the view that the logical form of a scientific theory is the valid but FFT syllogism.

Examination of the fictionalist view

(*a*) The general method of science is the discovery of causes through an examination of their effects. The application of fictionalism in astronomy should lead to a general fictionalism.

(*b*) The most comprehensive and the most convenient theory has title to be considered representative of reality unless it is shown positively that it is not.

(*c*) There are physical causal laws which lead from what a theory claims to be the case to the phenomena, whereas the FFT syllogism is internally connected only formally.

4. *Scepticism about hypotheses*

The absurdity of claiming that no theoretical statement can really refer to and describe real things and processes can be mitigated by conceding that theoretical statements are unverifiable but by defending the putative realism of theories.

This position was adopted by Osiander in his apologetic preface to the *De Revolutionibus* of Copernicus.

(*a*) The revival of the theory of Gassendi, under the title of 'instrumentalism', has led to the view that theories be judged for their effectiveness as instruments of inference among actual statements.

(*b*) The sceptic leaves science in much the same state as the fictionalist, in that it does not make sense to ask *which* picture of the world is true.

5. *The examples and the philosophical theories compared*

(*a*) The theory of mechanical force seems to fit the phenomenalist theory well, since 'force' can be treated as meaning nothing but 'mass acceleration'.

(*b*) Geometrical optics seems to fit the fictionalist position well since it seems as if a light ray is just a fiction, while 'heat' seems to be a concept well adapted to the sceptical point of view, since even in its present energy guise it is difficult both to deny the possibility of its existence, and to conceive of how that existence might be proved.

6. Realism

The case of the virus seems to suggest that theoretical statements do sometimes refer to real things, whose existence is capable, it turns out, of demonstration.

(a) A realist theory of science distinguishes between 'reference' by which words refer to things, whether those things are present or not; and 'demonstration', by which people point to things, in their presence. The former suggests hypotheses of existence, the latter constitutes proofs of existence.

(b) Three realist principles

(i) Some theoretical terms refer to hypothetical entities.
(ii) Some hypothetical entities are candidates for existence.
(iii) Some candidates for existence exist.

These principles can be illustrated by the theories of disease, from 1600 to now.

7. No concept remains fixed in a definite epistemological status.

(a) There can be a change from fiction to putative existent by a change in general metaphysical climate.

(b) There can be a change from being a mere candidate for existence either to real thing (virus) or fiction (heart septum pores).

8. The settlement of existential questions depends upon instruments in part.

(a) Sense extending instruments can be used to prove the existence of hypothetical entities.

(b) Detecting instruments can be used to prove existence, within the framework of a theory. The proviso has to do with what is said to exist, not that something exists; e.g. the electroscope as a detector of the electric field.

9. Realism and phenomenalism are poles in the epistemology of science.

Realism seems suited to sciences like anatomy, physiology, and chemistry, while physics seems the archetypal positivist or phenomenalist science in its more fundamental reaches. This is because at the boundaries of knowledge the phenomena are the limits of the conceivable.

4
Metaphysical Theories

METAPHYSICS IS THE study of the most general categories within which we think. There have been several important metaphysical systems, i.e. systems of concepts expressing these most general categories, which have functioned in the organization of scientific knowledge. We shall reach some understanding of them by a gradual unfolding, through various stages, of the details of their structures. We shall avoid questions as to the origin of the categories, and of the exact manner in which they enter into our thinking, that is, we shall not discuss the general question of the origin and growth of concepts. The most general categories enter our thinking in various ways which include the way we perceive the world, the way language is organized, and the choice of concepts from among the linguistic possibilities with which to describe and theorize about what we perceive. The categories with which we shall be concerned will be Substance, Quality, and Relation. The categories are in a way of reflection of the kinds of question we can put to nature. The category of Substance is concerned with Things, and is a reflection of the questions that can be asked about Material Stuff and Individual Things, that is, with such questions as *what* a thing is, and *which* thing it is. Or, as one might say, the category of Substance is concerned in the classification and identification of things. Here is the Koh-i-noor, an individual thing, and it is a diamond, a form of the material carbon. Applying these categories both in their general and particular forms allows us to bring order into the world as we perceive it, which is complex both in structure and behaviour. Is the great diversity of things fundamental or are they really varieties of one basic material? Are the individual things that we identify in our ordinary world themselves colonies of individuals, and does this process of ordered subdivision go on indefinitely? Or are there ultimate, indivisible, but non-infinitesimal individuals, or atoms? Are the qualities with which things appear in perception their real qualities? Is the very idea of an unperceived quality a nonsensical idea? Such questions cannot be satisfactorily answered by a more careful scrutiny of the world. They are metaphysical questions, and are about the concepts to be used in understanding things and processes.

Consider the question of whether there exist ultimate individuals or atoms which are resistant to decomposition by any kind of method of analysis. Suppose that all known methods of analysis have been tried and that the results of using them yield a class of individuals which resist all our efforts to analyse them further. To be sure that these were atoms we would need to be sure that another, more powerful means of analysis *could not* be found. But how could this be known? To use the concept of 'atom' of whatever entities resist analysis is to commit oneself beyond what can be ascertained as a matter of fact. It is to commit oneself to a metaphysical position, that is to choose a concept and the system of concepts that go with it with which to deal with the world. It is not to assent to a hypothesis about a possible discovery. If one is a metaphysical atomist the analysis of one's atoms by someone else is not a refutation of one's position. It simply leads one to make a new identification of what it is to be atomic.

I want to introduce the subject of the metaphysics of science by looking at some of the particular options among possible concepts which might be exercised under the various categories, before discussing some of the great, articulated systems which have dominated science so far. Under the category of substance we have the concepts of a 'material' and of an 'individual'.

Material

It is obvious that many common materials are discontinuous, that is, made up of granules. Though sand and cement conglomerate into concrete when mixed with water, the sand retains its granularity. Water seems continuous as a material, because though it can be separated into drops, these seem continuous too in their material and coalesce, rather than conglomerate, into what seems to be a continuous material. Which is ultimate, continuity of material as water seems to show, or discontinuity of material like sand? There can be nothing experimental in the way we reach our decisions, they transcend experiment. Our decisions represent our acts of adherence to metaphysical systems. Arguments can be advanced for and against particular choices among the options. The incoherence of an ultimate discontinuity view might be made out by some such argument as follows: Were material ultimately discontinuous we should reach final granules which were not themselves conglomerations of granules. Their material would therefore have to be continuous within their boundaries, hence in the end material must be continuous.

We are familiar too with the technique by which several materials are shown to be actually different forms of only one basic constituent. Ice, water, and steam; charcoal, diamond, and graphite; these are familiar examples. Is it perhaps the case that all diversity of material is but the allotropy of a single, basic stuff? Is there one ultimate material or many? How would such a question be settled? Two quite different problems can be identified here. There is the problem of the understanding of the philosophical concept of 'substance', that to which all properties are attributed. This concept makes its way into our thinking in the following way: Consider a tennis ball. Both its whiteness and its spherical shape are attributes of the rubber object. But what of its rubberiness? Of what is that the attribute? It might be argued that there is a basic material, a prime matter, substance in general, to which all attributes belong, as whiteness and sphericity belong to the rubber and cotton of which the ball is made. Such an argument would lead one to think that the diversity of materials with which we are accustomed to deal arises by diversification of one ultimate material.

A different sort of progression towards a unified theory of materials appears in the sciences too. Diversity of materials is explained by treating the diversity as due to differences in structure among a small number of more basic materials. The chemists' doctrine of elements is an example of this. The diversity of materials is explained by the theory that each is composed of molecules which are different combinations of atoms of a smaller number of basic materials. Thus methane gas is a material whose molecules or least parts are formed by combination of four hydrogen atoms with one carbon atom, while octane is a material the molecules of which are formed by the combination of eighteen hydrogen atoms with eight carbon atoms. There are thousands of different hydrocarbons whose diversity can be explained as due both to the different proportions of hydrogen and carbon atoms that are combined in their molecules, and to differing arrangements in space of the constituent atoms. Thus a thousandfold multiplicity of materials is transformed into a thousandfold multiplicity of combinations of a mere duality of materials. Similarly, physicists have offered an account of the hundredfold diversity of the chemical elements as due, not to a hundredfold diversity of materials but to there being one hundred different arrangements of a mere threefold diversity of more basic materials, protons, neutrons, and electrons. Encouraged perhaps by the philosophical arguments for unity of material, and now almost by instinct, scientists seek to transfer diversity of materials into diversity of structure of a lesser diversity of materials. Perhaps they hope in the end to reduce all diversity to structural diversity of combinations of only one kind of ultimate stuff.

Common experience seems to present us in a very natural way with a profound and seemingly irreducible difference between space and material. It seems very obvious that there are gaps between pieces of stuff, in which there is no material. But we soon learn that air fills the spaces between walls and the gaps between solid objects. Which option shall we take for forming our ideas of the ultimate? Is the air a swarm of particles with empty space between? Is there empty space between the stars? Or is the world a plenum, full of material, only some of it rather light stuff which offers practically no resistance to the movement of heavier pieces?

Despite the fact that the existence of space within which things exist seems to be obvious, most of the intellectual history of mankind has been passed in a belief in the plenum. This is the theory that the world is full of material. 'Nature abhors a vacuum' is a slogan with a very long history of assent. Arguments for the plenum range from Descartes's queer contention that if there really is nothing between the walls of a vessel then the walls must be in contact, to the sophisticated space-warps of Einsteinian gravitational theory. Arguments for the void, for empty space appear in two ways. In one the concept of a void is developed as an empirical concept to explain the apparently unimpeded motions of the heavenly bodies. But this theory has difficulties. They arise from attempts to explain the action which one body exerts upon another at a distance, particularly the action of gravity and of electric and magnetic forces. These forces seem to be able to act across the void, to act at a distance. And this is a profound violation of the models of action which our common experience has given us. We can act upon things only by contact. Are there two kinds of action, action by contact and action at a distance? Every effort has been made in the last three hundred years to eliminate action at a distance as a fundamental mode of action. The concept of the field is the most profound that has yet been developed to handle the problem.

There is yet another intellectual pressure which leads away from the void towards a plenum. It is an aim of science to seek for a unity of the world, for every thing to be connected with every other and to believe that action is proliferated through the Universe. The world is *somehow* to be conceived of as one. The forces which act from one thing to another, act on everything. The powers of all bodies are, in a way, universal. No thing can exist in total isolation. We shall see this concept of the Unity of the World appearing in different forms. Do we discover that the world is a unity? No. This is a form that we impose upon knowledge of the Universe.

Individuals

Considering the nature of our ordinary experience it is not surprising that the concept of an individual should play as pervasive a role in our thinking and perceiving as does the concept of material. Individuals can have names, can be pointed to sometimes, can be referred to by name, or characteristic description. There seem to be three main kinds of individual possible, three subspecies of the metaphysical category of individual. As models for each of these three kinds I shall take for one a diamond, for the second a person, and for the third a flash of lightning. Each of these entities is an individual. They can be pointed to, referred to verbally, and they exist in space and time without occupying the whole universe. And most important of all each of them is such that where they are no other individual of the same kind can be. We can cite a general principle which governs the concept of an individual: *No two individuals of the same kind can be in the same place at the same time.* Is this a fact? Did someone long ago discover it? No. If two individuals are at the same place at the same time, they are necessarily different in kind. A pain is where the pin is, so a pain must be a different kind of individual from a pin. These remarks illustrate how our concept of an individual is to be used, rather than express matters of fact. Well then, the diamond, the person, and the flash of lightning are all, in their own kind, individuals. But a diamond would endure for ever if it were left alone, a person has a beginning and an end but endures for a time, while a flash of lightning is ephemeral, it is a momentary event. It happens, then it is gone. I shall call these Parmenidean, Aristotelian, and Heraclitean individuals respectively.

Parmenidean individuals

Parmenides, a Greek, seems to have held that the true world was permanent and unchanging, that real change was impossible, that nothing could really perish or come to be. Generation and corruption, indeed all the changes that we see must be illusions and mere appearances. In an unchanging world how is even the illusion of change possible? One way this might be possible is if the individuals in the world were Parmenidean, changeless, permanent entities, but were capable of entering into a variety of combinations, forming and reforming different structures. Then the atoms would be Parmenidean, but the changing organization of the Parmenidean atoms into temporary structures would lead to the appearance of change. The ordinary things in the world, which are certainly perishable and come into existence, would, on this view, have to be temporary conglomerations of permanent atoms. This idea is echoed in the Corpuscularian philosophy of the seventeenth century, to which we shall turn later.

Heraclitean individuals

Heraclitus, a near contemporary of Parmenides, is credited with the aphorism that you cannot step in the same river twice. He was evidently struck by the temporary nature and changefulness of the things in the actual world. Though little is known of the details of his views we shall appropriate his name for a truly ephemeral individual, something which does not endure at all, but exists only once and at an instant. Even a flash of lightning is not a perfect Heraclitean individual since it takes some time, though a very short time. A truly Heraclitean individual is such as the beginning or the end of a flash of lightning. Perhaps the semi-permanence of things is but an appearance due to a sequence of a multitude of closely similar events. We have already noted a very famous theory called phenomenalism which seems to imply a concept of individuals not unlike those of a Heraclitean theory. Remember that according to phenomenalism the basic elements of experience are momentary sensations, the statistical regularities among which it is supposed to be the task of science to discover. The sensation of the moment is an approximation to a Heraclitean individual, since it lasts for a very short time. Indeed, some philosophers have tried to reconstruct the ordinary world out of momentary sensations, by supposing that the semi-permanent things of ordinary experience are aggregates of actual and possible sensations. The basic elements of sensation, like coloured flashes and feelings of touch, have been called 'sense data' to emphasize their allegedly fundamental role in knowledge. There is no doubt that the Heraclitean conception has been a popular one.

Aristotelian individuals

Aristotle frequently spoke of things as having potentialities for change. He insisted that things were capable of becoming other than they were. He supposed that generation and corruption were truly fundamental processes in the world, and responsible for the creation and destruction of the actual entities with which the world is furnished. Most plants are generated anew each spring and summer and decay in the autumn and winter. Over longer time spans the same is true of animals and men, and, we may suppose, of everything we can recognize as an individual if we consider it in relation to a long enough time. Upon what, then, does the permanence of the world depend? For an Aristotelian it depends upon indestructible matter which passes from individual to individual.

A consequence of this is a further refinement and modification of the notion of an individual. Individuality may be preserved through some changes but not through all. An Aristotelian individual does not have to be static to remain the same individual. A man in his prime was a child and will be a dotard, and still be the same man. Before his conception he did not exist as himself, but his component atoms did exist, and will persist even into other men. Each human being has a good chance of incorporating an atom or two once deployed in Julius Caesar's body. There seems to be no general specification of the limits of individual change. Sometimes what counts in a judgement of whether an individual has persisted is the continued possession of some set of essential qualities, e.g. a sword and a ploughshare are two different individuals, though the one may have been made out of the other, that is, they used the same metal. Other kinds of individuals, like men, for instance, seem to depend upon bodily continuity in time and space, and can show great changes in outward qualities while remaining the same individual.

Science has sometimes pursued the Parmenidean individual as an ideal, and sometimes the Heraclitean individual. However, whichever serves as the model for the *basic* individuals of the world some element of each must be incorporated in our idea of the individuals of the actual world they constitute, since the real things of the world, be they diamonds or lightning flashes, are actually Aristotelian individuals. Atomism is the Parmenidean response, and phenomenalism is the response of the Heraclitean to the demand for a specification of basic and ultimate individuals. Perhaps a whole-hearted adherence to the Aristotelian idea as a specification for the fundamental concept of an individual can resolve the tension between the two extreme views of the nature of individuals. Perhaps the things in the world really do come into existence and pass away, but perhaps they endure for a while between, and perhaps the permanence of the world comes from their not all being created and passing away together.

Qualities

Materials and individuals manifest themselves in our experience as having certain qualities. Snow is white and cold, Mao is tall and poetic, the Koh-i-noor is scintillating and valuable. We depend upon qualities both to identify a material and to differentiate an individual. The possession of some qualities is essential to the nature of a thing, and determines what kind of thing it is. But things also have qualities which are inessential, that is, which could have been different without our having to say that there was a different kind of thing. When animals and plants are classified into species, genera, and so on, differences in the qualities of specimens serve to mark off species one from each other, the essential qualities making up the definition of the species.

Three major theories of qualities have dominated scientific thinking. In the Aristotelian system qualities were treated as the actualizations of Forms in Matter. The model for this theory was perhaps something like a stonemason imposing sphericity upon a stone as he gradually chips and polishes away surplus material. Sphericity which was potential in the stone is slowly actualized. It is not easy to see this theory as so successful when dealing with warmth or colour. According to the theory a thing becoming hot is also to be seen as the gradual actualization of hotness in the matter of the thing. A thing is cold when the form of heat is little actualized. All the qualities things might have were treated along the same lines, the changes and acquisition of qualities being construed on the model of the artist and the artisan imposing form upon matter. This is an unfamiliar picture to us now but it dominated science from antiquity until the sixteenth century.

The rise of modern science was closely associated with the development of another theory, also known in antiquity. It is the theory of primary and secondary qualities. It has several forms, and I shall try to expound some sort of common denominator of the various theories. The theory maintains that many of the qualities that we suppose things and materials to possess in themselves, are not at all as we perceive them. The feeling we have of heat, our experience of the quality of warmth in a body, is quite unlike the quality that a body has when we feel it to be warm. The actual quality is a gentle motion of its components. We do not feel a motion, we feel warmth. Warmth is a secondary quality because warmth in a thing is actually a motion whereas warmth as we experience it is not. But some qualities are perceived pretty much as they really are in the thing which has them. Allowing for distortions due to point of view it remains true, according to this theory, that shapes are perceived as shapes, that motions which are perceived are really movements of things, that the number of things perceived is really the number of things there are. The qualities of which this is true are primary qualities.

The theory is completed by the principle that the primary qualities are the essential qualities of everything, that they are the real qualities of things, so that if we ask what warmth *really* is, its 'essence and quiddity', we learn from this theory that it is really a motion. In short, the secondary qualities are really primary qualities in a thing or material, only the secondary qualities manifest themselves in perception in a manner which is different from what they really are.

There is a third theory of qualities, hinted at in Locke, the chief advocate of the primary and secondary quality distinction. It is the idea that the qualities a thing or material manifests are the results of its powers. A thing is an agent which acts upon other things, bringing about changes in them, and this it does by virtue of its powers, the totality of which make up its essential nature. When a thing or material acts upon an organism it is manifested to the organism as having qualities, that is the organism sees it to be red or feels it to be rough. A thing or material is white because it has the power to reflect all the light which falls upon it, and when a thing of that kind is seen by a light-sensitive organism it is seen to be white. But to say of the thing that it is white is to say that it has the light-reflecting power, not that it has in itself, a certain quality of whiteness. And we know that it has such a power because it looks white.

Slowly the tide of scientific discovery has been flowing in the direction of the third of these theories. It now seems so clear that things and materials are the centres and bearers of powers, rather than the collocations of qualities or the actualized forms, that both the opposing theories seem implausible. Only perhaps by trying to enter into the Aristotelian or Corpuscularian world view can we recapture the state of mind within which either the forms theory or the sensory experiences theory of qualities makes sense. We shall see later, in some detail, how one of these theories of qualities, the theory of primary and secondary qualities, determined a system of description and explanation which survives into twentieth-century science, when the atomistic, Corpuscularian philosophy is dead. How far is science obstructed by dedication to ideal forms which belong properly to an antiquated way of thinking? This is very difficult to say in general. Discoveries of permanent value were made by Aristotelians and much of the science people learn in their school days is a summary of the discoveries of Corpuscularians. Neither the Aristotelian nor the Corpuscularian philosophy is wholly mistaken. The biological sciences are still dominated by the idea of development towards the actualization of a plan only latent or potential in the beginning of the life of an organism. Much of simple chemistry and physics is explicable as the combination and recombination of corpuscles, that is small particles. The Aristotelian view of the world is strained in trying to give an account of inorganic things and materials. The Corpuscularian philosophy offers a poor account of ultimate entities. But though the one is restricted and the other superficial they are both powerful conceptual tools and valuable specifications of the category of quality.

Relation

There is more than one individual thing in the world and more than one material. This fact is closely bound up in the theory of space and time with the way we have of individuating particular individuals by pointing to them. If there are two individuals and if they exist at the same time, then two simultaneous acts of pointing, one to each of them, must be possible. This possibility is created by their being separated in space, so that there are two places to which one can point simultaneously. Materials are not quite like this. Theoretically an individual could be just one point, and then two individuals would be two points and thus capable of being pointed out separately. An individual thing does not have to occupy a volume, a space, in order to exist. It just has to be at a place. But a material is different. A material could not be just a point. A material occupies a volume, it fills space. And two or more materials can be in the same volume, at least as far as our common experience goes. Salt can be dissolved in water, and then the salt and the water are in the same volume, in the same space.

However, so closely are the concepts of space, thing, and place intertwined that our system of concepts deviates from our ordinary experience so as to make it impossible for us to believe that two materials really are in the same place at the same time. We have to explain how it is possible for salt and water to occupy the same place, and our explanation consists in making each particulate. We suppose that both materials are made of corpuscles with gaps between, and into the gaps of the one collection of particles can go the particles of the other entity. We do not recognize genuine interpenetration of substance. But our system of concepts is not the only one that there has ever been. And we can easily imagine a system in which there was genuine interpenetration of material. Such interpenetration was recognized by Lamarck, who as well as being a great biologist, was a chemist of distinction, but who thought of chemical change as the gradual interpenetration of materials by each other, rather than as the rearrangement of atoms in space. Something of the Lamarckian conception of materials still exists in our ways of thinking. Two magnetic fields do not exclude each other but come together into a unified field, whose strength is a vector sum of the strengths of the original fields at each point. But the fields do genuinely interpenetrate and become one field. They are not separately identifiable within the final product.

The totality of possible pointings to possible contemporaneously existing things makes up space. Space is the totality of places where things can be at the same time. And if we think of the minimal thing as a simple point, then space is the totality of points.

With one further small step in thought, we shall have space as a system of relations. If we ask how the system of places where things can be is to be organized, we can start with a relation of great simplicity, namely that of 'betweenness'. If there are three things, then they can be arranged in space so that C is between A and B. This is a different configuration from that in which A is between B and C. The three things can be triply simultaneously pointed to, making three points, related by the relation of betweenness. Continuing in this and similar ways we can organize all possible places for pointing into a system of points, which can be considered and studied in abstraction from any particular things occupying the places in it. This system of points, related by such relations as betweenness, is space. To say that an individual thing is in space implies that it can be pointed to, and that means that it must be at one of the points of space.

Just as the concepts of 'material' and 'individual' are intertwined with the concept of 'space', so are events with times. Events and processes, changes, generation, and decay, do not just happen, they take time. What does this mean? Can two different events occur simultaneously at the same place? Only if they are of different kinds. A piece of toast can become both black and hot at the same time. But what cannot happen at the same time is that the toast should become both burnt and perfectly cooked. A banana cannot be both ripe and green at once. So if it is true of some particular banana that is green and that it is ripe, these facts must be true of the banana at *different times*. Time can in fact be thought of as just the set of changes in all the things in the world which exclude each other. To say that a happening is in time is just to say that it is a change of a thing or material from one state to another which excludes it. The totality of such changes *is* time. But time is also an organized system of moments. Ignoring the complications that arise from difficulties about finding out what is happening at other places and other times which are dealt with by the theory of relativity, which is an epistemological theory, as we have come to use that expression earlier on, we can do a great deal by way of organizing happenings by means of the relations 'before' (and 'not before' = 'after') and 'at the same time as'. If two incompatible happenings occur to a thing, then one is either before the other or after it; they cannot be simultaneous. But which is before and which after? Upon what is the direction of time based?

This is a topic fraught with immense difficulty. Some writers have tried to tie the direction of time to some great cosmic process. They have made it a sort of matter of fact that processes start in the past and finish in the future, with respect to some part of the process itself. Others have felt that it is a conceptual matter that the start of a process comes before the middle of it and that its end is after its beginning. They have felt that such remarks are not matters of fact at all but notes about the way the concepts of 'end', 'start', 'before', 'after', and so on are used, and connected with each other. For people who think along these lines it is not just tremendously difficult to imagine time travel, i.e. someone alive at present 'visiting' an event in the past, but it is logically impossible that there should be such a process. For a time traveller to encounter a happening from his past, or that had happened at a time past to him would be a logical impossibility, since the notion that that event had already happened follows from the admission that it was past, and to re-encounter it could only be to experience an event exactly like it but numerically different from it. If a time traveller encounters the Battle of Hastings, and that battle is past with respect to his birth and existence, then he cannot encounter the same Battle of Hastings at which Harold was struck in the eye by an arrow; but if he *thinks* he is, he must be encountering another battle exactly like it. The individuality of an event, in short, consists of its having happened.

Let us leave this problem, and suppose that we can tell which events come before which and that this order is unchangeable. Then all the events in the Universe can be ordered with respect to each other by the relations 'at the same time as' and 'before' (or 'after'), provided we ignore complications due to the fact that the signals by which we become acquainted with distant events take time to reach us.

By considering the events of the history of the world abstractly, say by numbering outstanding events, like the days, and the years, we can devise a system of time by which other happenings can be ordered, and which constitutes the basis of a systematic vocabulary for the identification and statement of when a happening takes place. Time does differ from space in one essential respect. While there can be places at which nothing is at some time, and there can be acts of pointing which fail to locate a thing or material, in short while there can be space empty of particular things or materials, there cannot be empty time. Since time just does consist of events, and happenings, and since in a certain sense, even if nothing is happening, the mere existence of the world is itself a happening, empty time cannot be. This is not a fact like the fact that arsenic cannot be an exclusive diet, or that a perpetual motion machine cannot exist. Rather, it is a logical fact. It follows from the nature of the concepts of 'time' and 'happening'. It is a fact like the fact that there cannot be a whole number between 4 and 5.

By referring individual things and materials to space, and happenings and processes to time, the relations between things can be expressed by the relations between their places, and the relations between happenings by reference to some standard process like a clock. In these ways structure can be expressed. To say that the atoms of hydrogen are at the vertices of a tetrahedron of which a carbon atom is the centre, expresses the structure of methane, by stating what are the spatial relations between its constituent atoms. The dating of the geological ages, Carboniferous, Pleistocene, and so on, similarly expresses the structure of the geological process. A great many important facts are facts of structure, and the spatio-temporal system provides us with a way of expressing those structural facts in terms of a universal and abstract system of relations.

This is an immense subject, and I have skated over it very lightly indeed. But enough has been said, I hope, to show the importance of the topic and the indispensable nature of the spatio-temporal relations in getting a grasp of a world in which there are a multitude of things and a great diversity of materials, and an ever-changing flux of happenings.

These relations of space and time are not relations of connection between things and happenings. A rope tied between two trees connects them, and if the one is pulled to one side the existence of the connection can transfer the effect to the other. But a spatial relation exists even between empty places, so the mere existence of spatial relations does not serve to bring things into connection. Things related only in space are independent. They can have no effect upon each other. Happenings related only in time are not connected either. The lighting of a fuse and the later explosion of the charge are connected happenings, through the process by which the fuse burns along its length until the charge itself is ignited. Some things and some materials and some happenings are connected, while some seem to be pretty well independent of each other. At least some are sufficiently independent of each other for changes in them not to bring about discernible effects in other things. It may, in the end, be true that one cannot pick a daffodil without troubling a star, but the disturbance in the heavens is not discernible. Small boys can shout insults behind the back of a deaf old man without disconcerting him in the least. Some happenings in our common experience do seem to be unconnected. Others do not. The flux of happenings is riddled with consequences. Turn a switch and the light comes on. Eat a green apple and a stomach-ache will surely follow. Do that again, and you will surely be punished. Breach the dam, and out flows the water. What is more, many of these connections are pretty strong ones. Water just cannot be banked up, and without a dam wall it must run down into the valley below. If the light does not come on when the switch is flipped, there is something wrong with the system, because if the system is all right, the light must come on. Perhaps a good digestion can cope with a green apple, but no human digestive system can cope with a gramme of strychnine.

These connections are causal, and because they apply between events, happenings, and states, and through processes, they are causal relations. It is to a general discussion of that system of relations that I turn now.

Two very different metaphysical theories of causality have waxed and waned in public esteem. There are various ways of expressing the contrast between them. Both start from the fact that some sequences of events are recognized as causal sequences and some as random or accidental sequences. What is the basis and what are the consequences of the distinction? In explaining the existence of a state of some sort in an individual, or in explaining the occurrence of a happening, reference is made to some other happening which is supposed to cause the one in which we are interested. It is also usually the case that we only set about a causal inquiry when something out of the ordinary comes about. We do not usually set about asking for the causes of the common run of things. It is part of the task of science to make that step, to ask for the causes of the commonplace. Why does the moon continue to circle the earth so regularly? Why does a cold get better in about three days? Why do cows give birth only to calves? There are all sorts of practical reasons why we are usually only interested in the causes of the unusual, but for the metaphysics of causality we must seek a theory which deals both with the causality of the unexpected and with the causality of the normal and the apparently accidental. Freud was a great scientist because he looked for the causes of such commonplace occurrences as slips of the tongue, as well as for the causes of such unusual happenings as fits of hysteria.

Not only do cause and effect constitute some sort of pair, but the cause is that member of the pair which comes first in time, or if it is causality between processes that we are studying, the process which is the cause is running contemporaneously with the effect and not after it. A cause cannot come later than its effect. Is this because we have never as a matter of fact found such a case? No, it is part of the metaphysics of causality, it is a consequence of the way we have constructed the concept of causality as an intellectual tool for dealing with the manifold and complex happenings of the world.

The two great metaphysical theories of causality take their difference from the way they treat the relation between the cause and its effect. In the *generative* theory the cause is supposed to have the power to generate the effect and is connected to it. In the *succession* theory a cause is just what usually comes before an event or state, and which comes to be called its cause because we acquire a psychological propensity to expect that kind of effect after the cause. The difference between the theories can be expressed in other ways. For the generative theory the relation between the events or states or happenings that are related causally is internal to them, the cause and the effect are not independent of each other, and the effect could not happen without the cause. It would not be just what it is were it differently caused. And part of what it is to be the happening or event which is a cause is to be just that happening which generates a certain effect. On the other hand, the succession theory treats the causal relation as external to the cause and effect so related. The cause can be described perfectly and completely without reference to what effects it has, and the effect of a cause is an independently specifiable event or happening or state which would be just what it is had it been spontaneously generated.

But the most famous way of making the distinction between the two great theories of causality is based upon the analysis of the concept of a causal 'connection'. The generative theory holds that there is a real connection between causes and their effects, and that in many cases this can be identified with a causal mechanism which on being stimulated by the cause produces the effect. For instance, if we are interested in the causes of cancer, the generative theory holds that the causal story is not completed by the citation of the statistics which show that heavy smoking is pretty regularly followed by cancer of the lung; it demands that this external relation between the process of smoking and the process of carcinogenesis be supplemented by an account in biochemical terms of the mechanism which acts between the stimulus of inhalation of smoke and genesis of the cysts and destructive tissue of the malignant cancer. The succession theory finds nothing empirical to answer to the connection between cause and effect.

Both are supposed to be happenings, and the successionist looks in vain for another happening which is the connection between them. Baffled in his attempt to observe the connection as a happening intermediate between cause and effect, he then turns to a psychological account of the idea of the connection between cause and effect. He finds this in habit. According to Hume, the great successionist, when we have perceived a regular succession between one kind of happening and another kind many times then we form a kind of mental habit, and come to expect the one on the occurrence of the other. We pronounce them, he says, to be connected.[1] But this idea of connection is no other than the result of a psychological phenomenon, the expectation based on habit, and is not an empirical concept applicable to the real world at all. In the real world, on this view, there is nothing but successions of happenings, some of which are repetitions of similar kinds of happening, and it is this similarity in repetition that we pick out with the notion of cause. It is a further consequence of this view that the happenings which are the causes and effects in the successionist view are independent of each other; so after one kind of event or happening has occurred, any other kind whatever is possible. It is logically fallacious, on this view, to infer from our present knowledge of the world how the world will be in the future. This generates the famous philosophical problem of induction to which we have already had occasion to refer.

On the generative view, however, given a certain happening in the world, and the world being what it is, replete with generative mechanisms located in the many things and materials that exist in nature, not every possible outcome is equally likely. Given the structure of the human body, its chemistry and physiology, which together constitute a complex of generative mechanisms of great sensitivity, the ingestion of a pint of alcohol is certainly not as likely to followed by sobriety as by drunkenness. Given the nature of calcium chloride and of sodium carbonate, the mixing of their solutions is bound to lead to a double decomposition and the precipitation of calcium carbonate. This is not because we have the statistics of huge numbers of cases where these sequences of events were observed, and that we have formed a strong habit of mind to expect these outcomes, rather it is because we have

[1] D. Hume, *An Enquiry Concerning Human Understanding*, ed. L. A. Selby-Bigge, 2nd edn. (Oxford: Clarendon Press, 1902), §§ 48–61.

some idea of the nature of the entities involved, and how they behave. It is our knowledge of the mechanism of chemical reaction that leads to our confidence in the behaviour of the two solutions when mixed. Of course in the alcohol case we do come to our beliefs through experience, but once these beliefs have been acquired, and the statistics and correlations of happenings and states obtained, the successionist can offer us no further task of unravelling to perform. For the believer in generative causality the existence of the statistics is but the first step in a long process of investigation which ends only when the nature of the things involved has been found out and the reasons for the statistics thus elicited. Science follows the generative rather than the successionist theory of causality.

The discovery of the mechanisms by which causes produce or generate their effects is a central part of a scientific investigation. The discoveries of the mechanism of chemical reactions, of the mechanism of inheritance, and so many more, are examples of the fulfilment of this search. But a word of caution is needed here as to the meaning of 'mechanism'. In ordinary English this word has two distinct meanings. Sometimes it means mechanical contrivance, a device that works with rigid connections, like levers, the intermeshing teeth of gears, axles, and strings. Sometimes it means something much more general, namely any kind of connection through which causes are effective. It is in the latter sense that I mean the word in this paragraph and throughout this book. It is in the latter sense that the word is used in science generally, in such diverse expressions as the mechanism of the distribution of seeds and the mechanism of star formation. In hardly any of these cases is any mechanical contrivance being referred to. So we must firmly grasp the idea that not all mechanisms are mechanical.

To sum up; in the absence of complete knowledge of a phenomenon, scientists will settle for the statistics of the conditions of its occurrence, rather than go on in total ignorance. But it is an aim of scientific explanation not only to know how things happen and in what order, that is, to know the laws of nature, but to know why they happen as they do, that is, to understand the natures of things and processes so that it can be seen why those laws of nature which we have discovered have the content and form they do. To make sure that this is clear let me offer one last example of this drawn from the previous discussion of epistemology. Mendel discovered a series of laws of nature governing the inheritance of characters from one generation to another. These were statistical and 'external' in character, telling us how the characteristics are distributed in the generations; but that was all. Such a discovery, important though it is, does not satisfy the demand for explanation. That is satisfied when, by careful study of the process of reproduction and the mechanism of inheritance, the system of genes and replicating molecular helices of atoms is finally uncovered. Only then is the explanation complete. In some famous cases the mechanism of a law may still elude us, hundreds of years after the discovery of the law. I think it is fair to say that though we know the Law of Mutual Gravitation extremely well, we have very little idea as to the mechanism of gravitational attraction.

Causes do not act in isolation. Causes act in a world replete, as I have emphasized, with causal mechanisms. Or, to put this point rather more generally, causes always act against a background of more or less permanent conditions. We have already noticed the idea, derived from human situations, that causes are only sought for the unusual. In human situations the causes of something are those changes in a fairly stable situation that are believed to lead to an unwelcome or surprising or illegal happening or state of affairs. This idea can be generalized and to a certain extent transfers to science. While scientists are concerned to discover and describe the causal mechanisms which from time to time can be stimulated to make particular things happen, finding out what these are requires experiment. In an experiment a typical procedure requires the stabilization of the conditions and the changing of only one relevant factor. This change then becomes the stimulus and the degree to which this factor is relevant to the total outcome in nature can be determined.

In nature nearly everything that happens is the outcome of a multiplicity of influences bearing upon a complex of causal mechanisms. Consider, for example, the investigation of the factors necessary for germination of seeds, a rudimentary but not untypical scientific investigation. The seed, and the organism which subsequently develops from it, is the mechanism which acts upon various ingredients which it draws from the air and the soil, reorganizing them into a form suitable to be incorporated within its own substance. By varying the temperature in which the development occurs, but keeping light, moisture, oxygen, and carbon dioxide and mineral conditions steady, the way in which temperature contributes to the development of the plant can be determined. And so on for other factors like light and moisture. Nowadays there exist techniques for separating out the different effects of different factors even in cases where many factors are acting together, in some special cases, but the method of isolating factors and stabilizing conditions remains a useful one. It depends on the distinction between causes and conditions, and on the fact that whatever factor is not stabilized but allowed to vary becomes, for the moment, *the* cause.

The laws of nature, which describe, as we have seen, how the mechanisms of nature respond to stimuli, must therefore describe possible causes, because they can guide us in our expectations as to which responses we should look for when a mechanism has been stimulated in a certain way, assuming that background conditions remain stable.

Materials and things are distinguished not only by their manifest qualities but also by their causal powers, which depend upon their internal constitutions. A material which is said to be inflammable, explosive, poisonous, or sweet, is described thus because it is likely to burn when ignited, to explode when detonated, to cause sickness and death when ingested, or to taste sweet when placed upon the tongue. But to say of such materials as petrol, dynamite, strychnine, or sugar that they are respectively inflammable, explosive, poisonous, and sweet is to say more than just what they will do or be or effect under certain circumstances: it is to say that they are of such a nature *now* that they *will* do these things under the appropriate circumstances. To be an inflammable substance, the substance must have such a chemical and physical constitution that it will burn when ignited.

Another important distinction to bear in mind is that between coming to possess or losing a quality, becoming red or being squashed flat, and continually possessing a quality, for example remaining cubical in shape. I have been using the word 'happening' to refer to changes in qualities and constitutions of, or relations between, things and materials; and the word 'state' to refer to the continued possession of some quality, constitution, power, capacity, or relation. Now bearing in mind the distinction between powers and qualities, and between happenings and states, I can make some general remarks about the kind of entities that that have causal relation one with another.

Happenings can certainly cause happenings. The impact of a falling stone causes the crack in the paving. And after the paving has cracked it then is in a certain more or less permanent state, and we can also say of that state of being cracked that the dropping of the stone caused it. So both happenings and states can be effects. I do not think that our conceptual system employs the notion of cause in such a say that a state can be a cause. We have already noticed how the cause is contrasted with the conditions in which it acts, and the difference between the cause and the conditions in which it acts is exactly the difference between happenings and states. Conditions are more or less permanent states. States, from the point of view of the active, causal end of the causal relation, are what constitute the conditions or part of the conditions under which a cause acts. If hydrogen and oxygen are put into a eudiometer and sparked then the mixture burns in a short flash. The cause is the spark, but unless the materials are in the appropriate state prior to the spark the explosion will not occur. One must keep one's powder dry.

The successionist view of causality and the generative view differ most widely over whether they admit causal powers or agents into their view of the world. On the successionist view things are passive and effects are what happen to them, brought about by influences from outside. The model for a causal relation is the impact of chunks of matter upon other chunks of matter. Suppose there is a football lying in the middle of the field. Unless something acts upon it there it will lie. A foot, or perhaps better for this model, another ball, bumping into it, propels it forward. Causes are never, on the successionist view, acting from within a thing, but are rather stimuli from without. The chain of external stimuli can be traced back indefinitely on this view. One ball hits another and so there is an effect, but that ball was put into motion by another stimulus, and so on.

The generative view sees materials and individual things as having causal powers which can be evoked in suitable circumstances. It sees the world rather on the model of the way an explosive can lie quietly dormant unless the circumstances are so arranged that the awful power of the dynamite is evoked by detonation. Dynamite is as quiescent as the football until it is stimulated, but the explosion which is the effect is not wholly due to the detonation, while the motion of the football is wholly due to that which strikes it. The explosion is due to the power of the dynamite. And it has that power in virtue of its chemical nature. The generative view sees the world rather on the model of springs wound up ready to unwind, of weights poised over drops ready, in virtue of their position, to crash to the ground with the slightest touch.

If we were to try to resolve the differences between the two ways of viewing the causal structure of the world and of understanding the causal relation we should have to argue that those cases which seem to be best described, say under the successionist view, can be equally well described under the generative view, or vice versa. I do not propose to do that here, since my purpose is no more than to try to make clear the different concepts of causality which can be developed to try to make sense of the happenings, states, and processes of the world.

It is a very important feature of our concept of causality and of our concepts of space and time, that changes in the spatial relations of individual things or of materials can never be causes, but can only be effects. Suppose that something does change when it is moved. How is this explained? Suppose that it is a small compass needle mounted on a pivot. This is moved and it is noticed that the needle swings round and after it has been moved two inches sets in a new direction. Do we explain this change as due to the movement in space? No, because our system of thought requires that we look elsewhere than to spatial change alone for a cause of the change. We look for some other individual thing or material near by which could effect the change, by coming to be differently related to the magnetic compass needle because of a change of place. It will be no surprise to find a magnet near the compass needle, and immediately we put down the change in the direction of the needle to a change in the relation of the compass and the magnet. On the other hand, changes in spatial relations can be an effect of a certain action. A force may be exerted on an individual thing solely to change its spatial relations with other things. If we encounter a large stone in the middle of the road and push and shove it out of the way, then our exerting of the forces upon it has the effect of changing the spatial relations of the stone with the edges of the road. Of course that change will not be the only effect of pushing the stone aside. Pressure will be exerted on the parts of the road that lie under the path of the stone as it is pushed aside and pressure will be relieved on that part where the stone formerly lay. These changes will bring about changes in the surface of the road and indeed below it. Sounds of scraping will be produced and so on.

The temporal order is completely immune from causality. Changes in time are neither causes nor effects. Though in legal language we can speak of the 'effluxion of time' as a cause, this is only a metaphor from a scientific point of view. If nothing acts upon a material or upon an individual thing, then nothing happens. But this does not prevent the stable substance and the quiescent individual persisting through time. Nor can temporal relations be changed by any action or activity or stimulus. It is not possible to make an event future that was past, or to change the order in which a pair of happenings occurred.

In our system of thought we find it convenient to distinguish between two kinds of causality, that in which changes occur in some individual or individuals, and that in which an individual is created. We have already seen that the world with which science deals is a world in which the predominant kind of things are Aristotelian individuals, that is things which come into existence and persist for a while and then come to an end. Metaphysicians may, as we have also seen, recommend adopting a different view of the ultimate constituents of things and materials. It may be either that of the permanent Parmenidean atom, or the fleeting Heraclitean entity. But the world with which we have to deal offers us no Parmenidean individuals and those which seem to be Heraclitean, like flashes of lightning, are found, on more detailed study, to be really rather short-lived Aristotelian ones. So we recognize changes in relatively permanent things, and creation and annihilation of things.

Creation, however, also offers alternative modes. We can conceive of creation in such a way that an individual suddenly comes to exist where there was nothing whatever before. This is creation from nothing, creation *ex nihilo*. An old slogan says *Ex nihilo nihil fit*, 'From nothing nothing comes.' This slogan denies the mode of creation by which something comes to be where nothing was before. For most cases of creation with which we have to deal there is no doubt that the slogan catches the truth of the matter. In our world creation is not from nothing but by the reorganization of what was already there. A new plant comes into existence by building up a structure out of elements drawn from air, earth, and water according to a plan derived from its parent plants. The same goes for animals. Chemical compounds of a novel kind are new arrangements of previously existing atoms of the permanent elements. A new house is made of bricks and mortar and these are themselves nothing but the processed materials from quarries and mines. In our world what is created is new forms, not new matter. The creation of new matter is not, I think, inconceivable. Indeed, some cosmologists have sought to explain the apparently fairly constant density of the material of the world and the apparently constantly expanding distances between the individual things by supposing that there is a constant and steady creation of matter from nothing material. New material is supposed to come into existence where previously there was none. Whether this happens is one thing, whether it is conceivable is another. And I think it can scarcely be denied that it is conceivable. Of course this creation will not be explicable, since it will have no antecedents by which it might be explained.

Whole libraries have been written in defence and explanation of the metaphysical options I have been so very briefly sketching. But I hope enough has been said to make clear what the leading motifs of each of the main ways of applying and specifying the categories has been. But, as I have already mentioned, these metaphysical matters may be looked at not only as options under each of the main categories by which the things and happenings in the world can be grouped, described, and understood, but may also be viewed as systems of concepts, that is, as three main world views. Since its beginnings in antiquity scientific knowledge has been articulated under one of three main metaphysical systems, which have served to provide, for each epoch, the forms of explanation. The three systems under which scientific mankind has so far lived are the Aristotelian system, the Corpuscularian philosophy, and the theory of the Plenum. They pictured the world respectively as Matter differentiated by Forms, as Moving Atoms in the Void, as a Universal Field in various and changing States of Strain. The first system dominated scientific thought from antiquity until the seventeenth century, at which time the Corpuscularian philosophy slowly displaced it, growing up within the ruins of the theory of Matter and Form. From about the middle of the nineteenth century the Corpuscularian Philosophy has been slowly decaying, though with a stronger and more conservative scientific establishment than existed in the seventeenth century the Corpuscularian philosophy has taken much longer to be finally displaced by a comprehensive theory of the field.

I have already briefly described the Aristotelian theory. I shall elaborate that description a little further, so as to contrast it a little more sharply with the Corpuscularian philosophy which displaced it. According to the Aristotelian view there is a primary and universal matter, of which everything that exists is made. This matter is differentiated into four primary elements, Earth, Air, Fire, and Water. This division of matter might be better expressed by saying that the primary matter is differentiated into those stuffs which are solid, those which are gaseous, those which are liquid, and those which are radiant. With the four elements go four primary qualities, or perhaps it would be better to say four basic natures, Hot, Cold, Wet, and Dry. For example, those materials which are naturally in the solid or earthy state are of a cold and dry nature, while those materials which are of a liquid nature are cold and wet, those of a gaseous nature are hot and wet, and those of a fiery or radiant nature are hot and dry. The exact nature of a thing would be determined by the proportions in which the basic natures were present in it, and that in turn would be determined by the proportion of the elements in the composition of a thing.

In medieval and later times alchemists adapted this system to their problems, and derived from Geber the idea of analysing actual materials by finding the proportions of the primary elements in their composition. From the numerological tradition of magic squares and gnomons, alchemists derived a numerical proportion of the elements which they believed would be the most perfect. Had they achieved the balance of elements indicated by this ratio they believed that they would have produced gold.

Complementary to this general theory of matter differentiated into four kinds and distinguished by four main qualities, natures, or forms, was a more particular, and in the seventeenth century more important, theory of forms. Aristotle had conceived of the changes in the world as due to processes by which what was potential became actual; what was less perfect strove for perfection; and what was not where it would naturally be strove to return to its natural place. Rather as shapeless lumps of clay are given a form by the potter, so the possession of any definite quality, whether it be a shape, colour, texture, medicinal power, and so on, is to be explained as the giving of a form to a previously formless matter. And just as the potter only gradually imposes the final perfection of definite form upon the clay which then becomes truly a pot, so only gradually does the form become actual in matter. Forms are actualized by degrees. And a thing may come to be possessed of a form to a higher or lower degree. A thing may become hotter, or rounder, or rougher, by changing the degree to which the appropriate form is actualized in it.

These ideas of the nature of change led to what is for science the most important part of the system. From it derived the Aristotelian idea of cause and hence a form of explanation which dominated science from antiquity until the end of the Middle Ages. A scientist, confronted with an individual thing, or a material, or a process in which he was interested, looked at it, if he were an Aristotelian, in terms derived from the metaphysics we have just sketched. He looked at it in terms of Form and Matter, of Potentiality and Actuality, of a Perfection to which all processes strove. There were thus four questions to be answered in any investigation: What Matter was involved? What Forms were involved? What End was being sought, or towards what End was the system moving if it was inanimate? And, finally, what was responsible in the particular case under investigation for there being a thing, material, process, or whatever to investigate? This has traditionally and rather misleadingly been called Aristotle's Theory of the Four Causes. It is often said that Aristotelians sought for four causes: a Material Cause, a Formal Cause, an Efficient Cause, and a Final Cause. It is then usually remarked that modern science has reduced this fourfold quest to one since it is said we nowadays recognize only the efficient cause. This is, at best, a half truth as we shall see later.

But we are still not at the heart of the Aristotelian system, for there is a deeper idea yet, the concept of the essential nature of a thing. To grasp the purport of this concept it is necessary to distinguish between occurrences which happen of necessity and those which happen by chance. If a stone is released some distance above the ground, then it must of necessity descend towards the earth.

It is in the nature of the stone to seek its natural place, which is to be in contact with the earth. But that the stone in descending knocks an elderly philosopher on the head, is something which happens by chance, because it is not in the essential nature of a stone to strike elderly but absent-minded persons. We can formulate laws of nature only for what flows from the essential natures of things, not for what happens by chance. There is a law of nature which concerns the descent of gravitating bodies, but there is no law of nature which concerns the accidents to which the absent-minded or unlucky are subject, nor is there a law of nature as to the kind of damage likely to ensue from falling stones. There might be none at all, there might be a disaster, but what happens will not flow from the essential nature of a stone.

From all this a form of explanation developed. A scientific explanation was complete when whatever had to be accounted for was given a Formal, Material, Efficient, and Final Cause; in short, when each of the four questions concerning matter, form, mode, and direction of development had been answered. The world was conceived as a complex of processes occurring in things and materials which were thought of more on the model of the growth of plants and animals (and the production of works of art and artefacts) than on the model of the thrusts and bumps of inorganic lumps of non-living materials. *This is not to say that the Aristotelians conceived of everything as alive and every process as organic.* It is rather that they chose, or had chosen for them by Plato and Aristotle, the originators of this view of the world, certain paradigms or models, by analogy with which all processes were conceived. Thus, having described the essential nature of an individual or a material by identifying its matter and form, one could then see what must of necessity happen as that individual or substance was subjected to the stimuli of accidents in its commerce with the other materials and things of the world. Several major scientific works of permanent importance were carried out within this tradition. Aristotle's great works on animals, Theodoric's study of the rainbow, the kinematic studies of the Merton College physicists under the leadership of Bradwardine, were all works of permanent importance and influence. In each study the scientist concerned sought the four causes as his solution to his problem, and having found them rested satisfied, since having thus met his ideal of explanation his task was completed. The supersession of this ideal by that derived from the Corpuscularian philosophy was one of the main features of the intellectual developments of the seventeenth century.

Most of the main ideas of the Corpuscularians have already been brought up at appropriate places in the early parts of this chapter. Now I shall draw them together to show how their special ideal of explanation followed from the system. The Corpuscularians conceived of every thing as composed of smaller parts, or corpuscles, arranged in empty space, or the void. Some Corpuscularians were also atomists, that is some believed that the processes of analysis which divided materials and things into their component corpuscles would eventually yield indivisible corpuscles, *minima naturalia*. They adopted the theory of primary and secondary qualities as an essential part of their philosophy of nature, and regarded their corpuscles as wholly defined by their bulk (size), figure (shape), texture (mutual arrangement), and motion. Geometry, the science of shape, and mechanics, the science of motion and action by contact, became for the Corpuscularians the fundamental sciences. Materials were distinguished by the shape, size, arrangement, and density of the corpuscles of which they were composed. All changes were to be attributed to the reorganization of the corpuscles of which things and materials were made up. Most Corpuscularians did not conceive it possible that the corpuscles themselves could change, and if a particle did seem to change, since change could be nothing but rearrangement of the corpuscular components of a thing, then the supposed corpuscle must be a compound body. It seemed to follow from this kind of reasoning that even if corpuscles were not strictly atoms, that is, not decomposable by any method of analysis, at least they were atoms for all practical purposes. These were the views of such people as Newton, Boyle, Locke, Harriot, Galileo, Descartes, and many other lesser figures, such as Hooke.

The corpuscles have persisted through the intervening centuries undergoing many changes, but our molecules, chemical atoms, even our protons, neutrons, and electrons are direct descendants of the small lumps of matter of which the Corpuscularians conceived the world to consist. Shape has become of lesser importance, and properties unknown to the Corpuscularians such as electric charge have become important, but the essential idea has remained more or less unchanged.

It is not so easy now to grasp the ways in which the Corpuscularians thought that causal relations existed among the changes in their universe of particles. One mode of causation was clearly successionist and has indeed proved to be the model for that mode of causation ever since. This is the kind of causation that occurs by change in the state of motion of corpuscles which is brought about by their running into one another, by contact or by impulse. A stationary corpuscle can be made to move only by being struck by another moving one. Similarly, should two moving corpuscles meet, then they are likely to meet at an oblique angle and to be moving with different speeds, and under those circumstances each will be moving at a different speed and in a different direction after they have collided. The explanation of the state of motion of the system after it has undergone collisions is certainly dependent upon its state of motion before.

However, another factor does enter in, something more like a generative element, something which depends upon the nature of the corpuscles themselves and not what happens to them. Corpuscles differ intrinsically in how much bulk they have, in their quantity of matter or what we should now call 'mass'. Their inertia, the property they have to resist changes in motion and to persist in a motion once acquired, is dependent upon their relative bulk. So if a very heavy corpuscle is struck by a fast-moving one, then it will not be set moving so quickly as will a lighter stationary one. Thus what happens to a corpuscle that has been struck by a moving corpuscle depends partly on the stimulus to which it is subjected and partly upon its intrinsic nature, its mass. Inertia is the power to resist change in motion and this is a property which a body has intrinsically, at least according to the Corpuscularian philosophy. Later critics, for example Ernst Mach, have tried to make inertia too an extrinsic characteristic of things, but I shall not go into their arguments here.[1]

Since the only way in which one individual could affect any other was by colliding with it (the mystery of gravity excepted), the Corpuscularians sought to reduce all change, of whatever kind, to changes which could be brought about by mechanical action, by collision, which might either change the state of motion of a whole body, or of its component parts, for example making them vibrate faster or more slowly. Or the impact might actually

[1] See E. Mach, *The Science of Mechanics.*, ch. 2.

detach bits or cause the internal parts to be arranged differently. In short there can only be change of structure, or change of motion, so all change must ultimately be change of structure or change of motion. It follows therefore that all causes of change must actually be changes of structure or motion changes. This chimes in, of course, with the doctrine of primary and secondary qualities. To effect changes in a secondary quality, since the quality is really some combination of primary qualities in the thing itself, like the number of corpuscles and their arrangement, something will have to be done to either the bulk, figure, texture, or motion of the thing to be affected, or some change made in the combinations of these qualities.

In the seventeenth century much thought was given to the establishment of the Corpuscularian philosophy. It was not received without opposition both from Aristotelians, and from those like Van Helmont who, while in opposition to Aristotle, nevertheless held the view that the elements of nature had a spontaneity, a power of self-development, which it is clear the corpuscles wholly lacked. Their only intrinsic power was the passive one of resisting change in motion. They were quite without the power to be agents, that is centres of causation from which effects flowed without being themselves but the passive transmitters of impulses from yet a further source. It was felt necessary to produce arguments or 'proofs' of the Corpuscularian philosophy. There were those who tried to show that the theory was true by means of arguments, and there were those who tried to show that the theory was true as a matter of fact. Locke and Descartes were typical of those who argued for the plausibility, and even necessity of different versions of the Corpuscularian philosophy. Gassendi argued for the possible truth of the system, and advanced factual evidence for its actually being true; Boyle (and earlier Bacon), assuming that the Corpuscularian philosophy was a consistent system of thought, advanced empirical or factual considerations in favour of it.

Boyle's reasons derived largely from attempts to show that the whole gamut of qualitative changes which can occur in nature and in the laboratory are brought about by changes that are ultimately to be conceived of as changes in the bulk, figure, texture, and motion of the insensible parts of materials. Grinding glass is just a process for reducing the bulk of the pieces of the material somewhat, and when this reduction in size has been achieved a striking qualitative change is seen. The previously transparent glass has now become white and opaque. Boyle cites, too, the example of white of egg, which is a clear, watery fluid. If it is beaten with a whisk it shortly changes its qualities, becoming white and sticky. If operated upon by the speck or cicatricula in the egg the very same fluid becomes reorganized in its minuter parts and acquires in the chick such qualities as springiness in the tendons, and a yellow colour in the skin of the young bird. Boyle notes that certain medicinal qualities can be acquired in this way, notably the power to cure the falling sickness which he thinks is a property of magpie chicks. Again in the preparation of certain chemicals with characteristic tastes, like Glauber's Salt, the new preparation has quite different properties from the substances and materials from which it has been made. In every one of these cases, according to Boyle, no agency has been at work other than reorganization. No new substances or materials bearing the new properties have been added in any of the cases, so the origin of the qualities cannot be external to the materials and mechanism involved. The qualities come into existence during the process of the formation of beaten egg, of chicks, and of Glauber's Salt. Those processes themselves, taking the chemical preparation as a model, are seen by Boyle as involving nothing but the separation of parts and their re-combinations, such as are involved in the case of beating the egg by a purely mechanical action, which can lead to nothing but changes in the bulk, figure, texture, or motion of the parts of the material which is being affected.

Thus, whatever may be the nature of the effects which nature or science can achieve, their causes must one and all be mechanical. From this the Corpuscularian Ideal of Explanation is reached by a short step. A complete and final explanation for the possession of a quality, and for any kind of change in quality, be it a single change or be it a process involving many particular changes, is always achieved when a system of causes is offered for the effects which turn out to be changes in the arrangements or in states of motion or in both, of the most fundamental corpuscles of which anything is made up. Bacon, writing much earlier than Boyle, had already asserted that in this way the true natures of things and qualities can be found, and with that final scientific explanation achieved.

This system of concepts was sanctified by Newton and to this day still exerts a powerful influence upon our thought, since, as we shall see in the next chapter, much of the shape of the scientific method to which we still subscribe is derived from this way of thinking about the world.

The Corpuscularian philosophy, like its predecessor, the Aristotelian system, is a metaphysical system. It is a way of thinking about the world, and it works by persuading us, in advance of any particular experience, to adopt a certain way of specifying the categories of Substance, Quality, and Relation. Armed with these concepts we then approach the world. No startling new fact led to the overthrow of the Aristotelian system. It was not that a new instrument was suddenly devised by which the dance of the atoms became immediately visible. No such instrument exists to this day. Nor is it the case that someone suddenly devised an overwhelming argument which disposed of the Aristotelian system by showing an internal inconsistency within it and proved the necessity of the Corpuscularian philosophy. Ironically, an argument has existed since the end of the eighteenth century, due to Boscovich, and repeated in a less clear form by Kant and others, which *validly* accuses the Corpuscularian philosophy of a fundamental inconsistency, such that it cannot serve as an ultimate explanatory system. This argument is correct! But it did not affect the popularity of the Corpuscularian system as a provisional way of explaining happenings and processes, and it did not even stop many people from supposing the system fundamental. The unpopularity of the atomic conception of matter in the nineteenth century is to be put down to other causes again, and though none of them are really empirical, none of them are due to observations. This interesting chapter in the history of the metaphysics of science has been admirably treated by D. M. Knight in his excellent *Atoms and Elements*.[1]

The most that one can say, short of writing a treatise upon each particular shift from one metaphysical system to another, detailing all the influences (and they are nothing short of the whole fabric of life and culture), is that when one mode of thought begins to be increasingly hard to apply to certain discoveries, a ferment of ideas begins within the intellectual community. The community from which new ideas come is not confined to scientists. Some of

[1] D. M. Knight, op. cit., ch. 2 and ch. 6.

the new ideas seem to provide ways of thinking about specific problems that are a great advance, or seem to be a great advance on the awkward and complex conceptualizations of the previous system. How much easier it is to explain the operations of the heart and blood system as the circulation of a single fluid than as the complex ebb and flow of several. And, ironically, the new and ultimately easier explanation is often harder for people trained in the old way of thinking to take in. Piecemeal success in a wide range of different problems then seems to encourage syntheses, in which the ideas are brought together in a system. Finally a group of philosophers appear whose arguments weld the new concepts together into a coherent system, and give them something of the appearance of necessity.

We can write the history of the origin of the Corpuscularian philosophy and the replacement of the Aristotelian system in the three hundred years from 1500 to 1800. Our intellectual climate, though, is not Corpuscularian. Since the days of Faraday, who offered us another way of looking at the world, the atomic conception of nature has been slowly decaying. But our situation is different from that of the seventeenth century. Whereas in most of the processes to which the Aristotelian system of concepts had been applied the Corpuscularians were right to see particles in process of rearrangement and change of motion, they were wrong to our Faradayian way of thinking only in supposing that particles were ultimate. So we shall turn now to a discussion of a whole range of ideas, which powerfully influence our present ways of thinking about the world and understanding its processes; which are clearly derivative from the Corpuscularian philosophy; and which, because of the truth that many processes are, at a first stage of analysis, really Corpuscularian, are still valid in our world today.

Summary of the argument

1. Metaphysics is the study of the most general categories within which we think.

(a) The categories are reflections of the kinds of question that can be asked about the world.

(b) Problems about the choice of categories, such as substance and qualities, or individuals and powers, are not empirical questions, but are not meaningless and are capable of rational discussion.

(c) A particular system of categories can be adhered to, despite apparent empirical difficulties, e.g. an atomist can always maintain that there are ultimate indivisibles, whatever has been divided.

2. Materials

Each category is distinguished by variants which might be chosen, within it.

(a) Materials which come under the general category of substance can be discontinuous like sand, or continuous, like water. A metaphysical option is exercised when we decide upon whether all materials are ultimately continuous or discontinuous.

(b) Materials present an apparent diversity. Some can be shown empirically to be different forms of the same basic material. A metaphysical option is exercised in supposing all materials to be forms of only one basic material.

(c) It is an important methodological principle that diversity of materials is to be explained as due to a diversity of structure of fewer and simple materials, e.g. the principle which lies behind the atomic and molecular theories in chemistry.

(d) A further basic distinction is that between material and space. It is a metaphysical question as to whether this distinction is ultimate. But it determines the form of scientific theories, between action by contact, action at a distance, and field theories. Decisions here are partly determined by further metaphysical principles, such as the Principle of the Unity of the World.

3. Individuals

(a) This metaphysical category is related to logical principles governing the function of reference in language.

(b) The category is also related to the concepts of space and time through such principles as no two individuals of the same kind can be in the same place at the same time.

(c) There are three main kinds of individuals:

(i) Parmenidean individuals are changeless, permanent entities. Atomism derives from the idea that if the fundamental individuals are Parmenidean then change can only occur through their rearrangement.

(ii) Heraclitean individuals are ephemeral, fleeting entities, existing only for an instant. In practice no individuals are really Heraclitean, but endure for very short times, e.g. momentary sensations.

(iii) Aristotelian individuals are created and destroyed, and endure for a while. They are capable of some changes without loss of identity.

(d) Science has assumed that, though the given individuals in the world are Aristotelian, ultimate individuals are either Parmenidean (atomism) or Heraclitean (phenomenalism).

4. Qualities

Materials and individuals manifest themselves in our experience as having qualities.

(a) Qualities form the basis of identification of things and materials. The set of qualities required for this purpose is the nominal essence of the individual or material.

(b) There are three theories of qualities:

(i) Aristotelian qualities are the actualizations of forms in matter.

(ii) Primary qualities are those qualities we perceive in things and materials which they really have, e.g. shape, while secondary qualities are those qualities which are not a bit like the qualities we perceive as a result of their effect upon us, e.g. perceived heat is not like motion. The theory is developed as a basis for science by the assumption that the secondary qualities are some function of primary qualities.

(iii) Though a thing or material manifests itself qualitatively, the fact that it is not always being perceived demands that the *power* to manifest the quality should be ascribed to the thing or material.

Modern science uses the primary/secondary quality theory as the basis of a first level of analysis, and the powers theory as a basis for fundamental theories.

5. Relation

There is more than one thing and more than one material and more than one event, so the question of the relations between things, materials, and events arises.

(a) Space

(i) Created by the possibility of separate acts of reference to two co-existing things.

(ii) Materials are continuous over a space so defined.

(iii) It is necessarily true that two individuals cannot occupy the same place at the same time, but it seems to be an empirical fact that two materials can occupy the same volume at the same time, e.g. salt dissolved in water.

(iv) Our conceptual system, using a Corpuscularian first level analysis of materials into individuals, requires a theoretical explanation of the empirical fact in (iii), and achieves it by supposing that materials contain void spaces between their component parts. This is not the only possible solution. Lamarck proposed a theory of genuinely interpenetrating substance. We use just such a theory in the case of inter-penetrating fields.

(v) Geometry, the mathematics of space, derives from the use of the concept of 'between' together with principle (i) above.

(*b*) Time

(i) Created by the possibility of two true but incompatible predications of properties to the same place. The resolution is achieved by the theory that they do not exist together, and thus there are different times.

(ii) The concept of duration develops through the observation that some things and materials do not change when others do.

(iii) Chronometry, the mathematics of time, derives from the use of the concept of 'before' (or 'after').

(iv) Two deep metaphysical problems arise here:

(A) Is the order of events, the 'direction of time', a matter of logic and grammar, or a matter of fact?

(B) Is our actual inability to 'revisit' the past a consequence of the logical features of the concept of 'the past', or an empirical matter?

6. The relations of space and time are not relations of connection, and changes in them are always effects, *never causes*. This principle is a contingent feature of our conceptual system, not true of the Aristotelian system, and challenged to some degree by geometrical interpretations of General Relativity.

7. Causal Relations have been the subject of two incompatible metaphysical theories.

(*a*) In practice we seek only for the causes of the unexpected, but scientists also seek for the causes of the unexceptional, e.g. Newton, Freud, and so on.

(*b*) The generative theory of causality:
(i) The cause produces the effect by the working of some mechanism.
(ii) Given the cause it is naturally necessary that the effect will occur, i.e. the effect must occur unless something interferes.
(iii) the connection between cause and effect is real, and is the causal mechanism.

(*c*) The successionist theory of causality:
(i) The cause is simply that independent happening that is statistically associated with the effect in its vicinity.
(ii) Given the existence of the cause it is possible that any kind of event could follow.

(iii) The connection between cause and effect is a psychological phenomenon only.

(d) The successionist theory derives from the prior metaphysical theory that the world consists only of atomic events. From (c) (ii) above the problem of induction follows, i.e. there can be no rationally grounded knowledge of the future course of events.

(e) Science is based upon the generative theory, and treats the statistical evidence of succession as the basis for the hypothesis that a causal mechanism exists.

(f) This generates a methodological principle, in that a study is deemed complete only when the causal mechanism has been identified, e.g. studies of the causes of cancer.

(g) This also determines the structure of scientific explanation, which consists both of statistical relations between the phenomena and accounts of the causal mechanisms at work, e.g. the explanation of inheritance.

(h) The causal powers of a thing or material are related to what causal mechanisms it contains. These determine how it will react to stimuli.

(i) In contrast the successionist view conceives of things as wholly passive and having no power. The total explanation of change comes from without.

(j) Since causation is the reaction of permanent causal mechanisms, causes must be changes in the conditions in which the causal mechanism previously existed. But effects might be either subsequent changes or the new state produced by the reaction of the causal mechanism.

(k) Apparent production of effects by the change of spatial or temporal relations is explained by the influence of neighbouring things and materials, whose effects are distance or time dependent, e.g. magnetic fields, etc.

(l) The generative theory of causality admits no creation *ex nihilo*.

8. Since antiquity the metaphysical background of science has been articulated under three main systems of concepts.

(a) The Aristotelian system:
(i) Ultimately there is only one substance, differentiated into four main kinds of materials, solid, liquid, gaseous, and radiant. Ordinary things derive their properties from the proportions of each major kind of material in their composition.

(ii) By the end of the Middle Ages this theory had been supplemented by the theory of substantial forms, in which there were a multiplicity of forms seeking embodiment in matter.

(iii) There is a general tendency according to this theory, for things to develop towards their most perfect manifestation, and this provides the dynamics of change from the merely potential to the actual.

(iv) The theory led to a certain style of scientific investigation which sought answers to four questions:

(A) What materials are involved in the phenomenon?
(B) What forms are being actualized in the process?
(C) What is the perfection at which the process is aimed?
(D) What is the initial stimulus which sets it off?

This has been rather misleadingly described as Aristotle's theory of the four causes, material, formal, final, and efficient.

(v) Science is concerned only with those happenings which come about of necessity, and not from chance, i.e. those which flow from the essential natures of things, and thus can be understood rationally.

(vi) Important scientific work was done within this system, e.g. Aristotle's biology, Theodoric's study of the rainbow, the Merton College school of mathematical physics etc.

(b) The Corpuscularian system:

(i) Ultimately there is only one substance, and change is possible because it is divided into units, which are capable of motion and hence rearrangement.

(ii) The arrangements of the corpuscles is the real essence of bodies, and defines their primary qualities.

(iii) Geometry as the science of shape, and mechanics as the science of motion are the fundamental sciences.

(iv) Changes in our ideas of secondary qualities are the result of changes in the arrangement and state of motion of constituent corpuscles.

(v) Basically the only real happenings were redistributions of motion brought about through action by contact.

(vi) The ultimate properties of corpuscles were their power to fill space (their extension), and their power to resist instantaneous increments of motion (their inertia).

(vii) Proofs of the Corpuscularian philosophy were attempted by Boyle, particularly, using the principle that mechanical changes bring about changes only in the bulk, figure, arrangement, and motion of the parts of things, so if qualitative changes follow mechanical operations, ultimately those qualitative changes must have been brought about by changes in the primary qualities.

(viii) The attempt to develop a consistent system of mechanical action by contact among solid, incompressible bodies was shown by R. J. Boscovich to be incoherent and to involve contradictions.

(*c*) Changes in a conceptual system are brought about by a variety of factors, such as increasing complexity of explanations using the old system as basis, discovery of inconsistencies within the old system, changes in general outlook, and so on.

(*d*) The powers system :

(i) The concept of fields of potentials replaces that of moving corpuscles.

(ii) Our conceptual system derives from the conceptual necessities of understanding Faraday's researches into electricity, magnetism, and light.

5

The Corpuscularian Inheritance

THE CONSTRUCTIVE ROLE of philosophy of science has, I
hope, slowly become clear as we have wandered through the
thickets of the traditional divisions of philosophy, logic, epistemology, and metaphysics. The philosophy of science extracts and
makes explicit ideal forms towards which the forms of description
and explanation tend, under the pressures of the systems and
doctrines and theories in each of the traditional branches of philosophy. Though we no longer accept the Corpuscularian philosophy as an exclusive system of concepts with which to gain
intellectual comprehension of the world, it still exerts a powerful
influence on the ideals according to which we describe and explain
individual things, materials, and phenomena. It seems to me that
there survive three main conceptions, specified in various ways,
which clearly derive from the Corpuscularian philosophy.

The principle of structural explanation

First, the properties of individual things and of materials, that
is their powers and qualities, including the causes of their sensible
qualities, that is the qualities they manifest to us when we perceive
them, should be redefined for scientific purposes as structural relations among standard elementary individuals. The presence
of a certain mode of organization among the elementary parts of
a thing becomes the main feature of the explanation of the powers
of individual things and materials to manifest certain qualities to
us, and to affect other things. It is in exactly this spirit that Bacon

advocated the replacement for scientific purposes of 'X is hot' by 'There is a "motion, expansive, restrained, and acting in its strife upon the smaller particles" of X.' This definition was proffered at the turn of the sixteenth century.[1] It is still in this spirit that a contemporary scientist distinguishes between graphite, carbon black, and diamond by reference to the different ways in which the elementary carbon atoms are arranged in space in each material. For example the power of graphite to act as a lubricant is explained by the arrangement of its component atoms in planes, laid one over the other, in such a way that the forces holding the atoms together in a plane are much greater than forces acting between the planes, thus allowing the planes to slip freely over each other.

The effect of tacit subscription to this form of analysis and explanation is that certain modes of description and explanation tend to be preferred to others. And this manifests itself in the ways in which the multiplicity of kinds of individuals and materials known in nature are treated. There is a tendency to try to reduce the number of kinds of individuals to as few as possible, and to reduce as far as possible the multiplicity of materials. This tendency is fulfilled in a rather particular way. For example, the ninety-two different natural elements, ninety-two different basic materials, have been reduced in recent times from ninety-two independent and different materials. Their ninety-twofold diversity is explained by supposing that each is composed of units having ninety-two different arrangements in space, that is, different kinds of atoms are differently related systems of only three basic kinds of individuals, protons, neutrons, and electrons. Differences in the powers and qualities of materials are then attributed to differences in the structure of their elementary constituents, that is qualitative difference is reduced to, and explained by, difference in the relations between parts. There is, then, it seems an advance in knowledge, a movement towards deeper understanding, when ninety-two different ways of relating three fundamental kinds of things replaced ninety-two different 'basic' kinds of atom. Proliferation of kinds in some science, as the kinds of atoms known proliferated in the nineteenth century, is regarded by those who subscribe to the Corpuscularian ideal as setting a task. It poses a challenge, to show that the multiplicity is nothing but diversity of

[1] F. Bacon, *Novum Organum*, in *Works*, edited by J. Spedding, R. L. Ellis, and D. D. Heath (London, 1857–74), I, 390.

relations, and not diversity of substance, i.e. of basic kinds of materials.

The Aristotelian system, using matter and form as its fundamental concepts, did not, in practice, develop a reductive approach to diversity despite the fact that it included the idea of the Four Elements. Instead of treating the diversity of substances as due to different proportions of the Four Elements which would have been the natural move, given the theory, the Aristotelians tended in practice to accept the multiplicity of substances as irreducible, and to postulate a separate form for every different material. This does not seem to me to be in any way an essential feature of the Aristotelian system. But it was a feature of the system as the Corpuscularians knew it, and indeed it was one of the features which drew down their greatest scorn. In *The Origins of Forms and Qualities* and in other works, Boyle in particular attacked the way in which Aristotelians had allowed independent forms to proliferate at such a rate as to correspond to the discoveries of new materials.

In all the sciences, whether chemistry, physics, or biology, the idea of structure has been most potent. Chemists have successfully shown that in molecules the constituent atoms are arranged in a definite spatial configuration, and that the relations between the atoms are crucial in explaining many of the ways that compounds behave. Similarly, anatomy, the study of the spatial relations of the organs and cells of the bodies of plants and animals, has been pursued so far as to include the study of the 'anatomy' of the constituent molecules of organic materials and the fine structure of living individuals. In fact biology has increasingly shown that diversity of kinds is not to be put down to an ultimate diversity of materials but to the diversity of structure among quite a small number of kinds of atoms, that is, to quite a small number of kinds of individuals, that is, from quite a small number of basic materials.

The Corpuscularian ideal of structure is modified in practice by a complementary tendency to that of the reduction of kinds of materials and kinds of individual things. It is a tendency to try to reduce the variety of relations which are used to build up the structures which are held to be responsible for diversity of kinds. Our ideas on both structural and causal relations seem to be affected in this way. Newton's lifelong dissatisfaction with gravity is an example of this tendency, I think. He had already developed a comprehensive system of mechanics for the behaviour of individual things based upon the single relation of collision between moving bodies. His laws of impact describe the redistribution of velocities and changes in direction that take place when two or more things collide. This is the very essence of Action by Contact. In common with most Corpuscularians Newton certainly believed that most action was in the last analysis Action by Contact. In the end, then, there was to be only one causal relation between individual things and materials, the relation of contact during which the only possible kind of changes allowable in a Corpuscularian world could take place, namely change of the direction or degree of motion. Every change which was seen, felt, tasted, touched, or observed in any other way would be reduced to a change in the direction or degree of motion of the corpuscles involved.

But gravity defied this programme. Gravity certainly affected both degree and direction of motion, and yet was the antithesis to Action by Contact. It was paradigmatically action at a distance, a mode of causation which seemed to have no place in the Corpuscularian philosophy. Newton disliked this very much and was at pains to explain that he did not consider gravity to be a primary power of matter, since it acted at a distance and acted with different force depending upon the distance apart of the interacting bodies. He hoped that it would be shown that it was the effect of an 'Agent acting constantly'.[1] And from his remarks in letters it seems evident that some form of action by contact would have satisfied him, in that it would have removed the blemish which he felt in his system, of admitting *two* characteristic modes of causality between things and materials.

This Corpuscularian ideal, according to which qualities, powers, properties, materials, states, and processes are to be analysed, is associated with a certain mode of explanation. Since the idea of structure determines what we are to think there really is in the world, it will also determine what we take to be acceptable as an explanation. It will determine when explanation stops. All these notions, that of the nature of a final explanation, that of which kinds of states of material things are ultimately responsible for the states of the world as we perceive it, and our ideas of what

[1] Newton, *Papers and Letters on Natural Philosophy*, ed. I. B. Cohen (Cambridge: Cambridge University Press, and Cambridge, Mass.: Harvard University Press, 1958), p. 298.

things *really* are, are connected, and are specified by different applications of the very same idea. The explanation of isomerism is an excellent example of the way these notions interlock. Isomerism is the phenomenon in which the two or more substances which have the same chemical composition show markedly different chemical and physical properties, so different as to enjoin their being treated as different substances. The explanation of isomerism is that the atoms, whose proportions are identical in each isomer, are differently arranged in space, and that the difference between isomers is due entirely to the different orientation and arrangement of the constituent atoms. Our tacit adherence to the idea that materials are really structures shows itself in the fact that, differences in geometrical structure are fully *acceptable* as the true explanation of isomerism. Having given a structural explanation and shown how it works in particular cases, there is nothing more to do. Put this another way: the structural account gives an account of what isomers *really* are, and what the difference between them really is. Finally it is the difference in structure which is *responsible for*, and is *the cause of* the differences between the isomers.

Taking this Corpuscularian idea of structure as defining an ideal form of explanation invites us to explain the properties and powers of individual things and of materials as due to their fine structure, that is as due to the dispositions and interactions of their parts. We could call this *microexplanation*. The global properties of individuals become functions of the properties of their parts. The mass of a solid body is the sum of the masses of the parts into which it can be broken up. Mass is an additive function of micromasses. The total charge on a body is the totality of elementary charges on that body. Charge is an additive function of microcharges. It seems to be perfectly conceivable that there should be relations between global properties and microproperties that would not be additive, in which the global property was not just the sum of the microproperties. If one is trying to relate the behaviour of a gas sample considered as an aggregate of molecules to the properties, structure, and behaviour of the molecules themselves, then the macroproperties are not just additive functions of the microproperties Pressure is related to the microvelocities of the component molecules, but pressure is calculated from these in a very complex non-additive way. In fact the sample of gas has no velocity which is the additive sum of the microvelocities; indeed it is not even the vector sum of those velocities, since the sample of gas has no velocity at all. Again, should we be attempting to relate the behaviour of a crowd of people to the behaviour of its individual members, the characteristics of a crowd are not additive functions of the characteristics of its members. Many groups or aggregates have properties that are not properties of the individuals of which they are a collection. Such properties are called 'emergent' properties. Individual molecules do not have pressure or temperature, these are properties only of aggregates of molecules. Individual cells cannot think or talk, but the aggregate that is a human being can do so. Emergent properties are particularly prominent in biological sciences, but they are commonplace elsewhere too. There is nothing particularly mysterious about them. They derive from two separate features of aggregates.

The kind of aggregates with which a biologist is concerned are organisms, and organisms have a structure. They are such that the structure is preserved during the life of the organism, and if the structure is modified, there is a continuity of structure between the various forms that the organism may manifest during its life cycle. Many emergent properties of organisms are to be explained by characteristics of the structure, and not just by the components that enter into that structure. The temperature maintenance of a complex organism is an emergent character, because it derives from the way the parts of the organism are arranged and how they are related to each other causally. The temperature of a warm-blooded organism is not an additive function of its component temperatures, as its mass is an additive function of its component masses. Probably the powers of higher organisms to solve problems, and of the highest to talk, are to be put down to the way in which the parts of the organism are related, and so to its structure.

But there is another source of emergent properties. This is what one might call for want of a better word 'the fact of the ensemble'. A single cell could be characterized by the totality of the chemical reactions that take place in it, many of these being catalytic reactions, catalysed at low temperatures by enzymes. There is an ensemble of chemical reactions in the cell, and the whole life of the cell is determined by them and the order in which they occur. This is itself controlled by the genetic material which is ultimately responsible for the enzymes that are synthesized by the cell, and in what order they are produced. But the fact that a certain ensemble of reactions are present in the cell is not a chemical fact. The explanation of why a certain reaction takes place can be referred to the genetic material and the explanation of why it is coded to produce just that enzyme at that time can be referred to the evolutionary history of the organism. But neither the genetic fact nor the evolutionary fact which explains the genetic fact are chemical facts. The explanation of the exact structure of the chemistry of a cell is a non-chemical explanation. It is a biological explanation. With respect to the facts of the chemistry of the cell, the fact that that chemistry consists of a certain ensemble of reactions, is an emergent property of the cell. It cannot be expressed in a list of chemical facts, however long.

But explanation does not have to be microexplanation. There is no absolute necessity about the Corpuscularian point of view. In our system of thought we admit macroexplanations as well, that is explanations of the nature and structure of the parts of an individual thing in terms of the characteristics of the whole thing. This is extremely common in everyday explanations. For example this is how social roles are explained. The duties of the chairman are explained by reference to the powers and functions of the committee, which is the collection of members, including the chairman. 'The chairman has a casting vote in order to prevent deadlocks', is a macroexplanation of his special voting power since it is the committee that fails to come to a decision, not any individual members of it. In anthropology the functionalist approach depends heavily on the use of macroexplanations. In that theory every social phenomenon is understood by reference to the function that it performs in the life of the society. So the explanation of the custom of giving gourds to the chief, who after making a kind of throne of them, distributes them to certain people, is to be found in the role that this performance has in the economic and ceremonial life of the community.

There are macroexplanations in physics. It may be true that some feature of the whole endows the parts with properties. Such a macroexplanation is found in Mach's theory of mass, according to which the mass of each individual body in the Universe is due to the effects of all the other bodies upon it. In a universe in which a body was alone it would have no inertia. So if we want to explain why individual bodies have inertia we look for an explanation in some characteristic of the system within which that body is a part.

In the first example we looked at, the move towards a preferred form of explanation began as a demand in favour of a certain kind of description, and gradually we saw it become a demand in favour of a certain kind of explanation. Certain kinds of causes were favoured, and explanations come of causes. The principle I am going to discuss makes a more purely descriptive demand. It is this: qualitative descriptions should be replaced wherever possible by quantitative descriptions. To elucidate this principle I shall show it operating in three different manifestations.

The first manifestation I want to discuss is the preference we have for the use of standard units in measurement over simple ordering scales. We prefer, for example, to use units of length rather than simple comparisons of length. An example would be the use of the metre to compare two swimming pools, one 50 m against the 100 m of the other, as opposed to saying vaguely that the second is much longer than the first. The use of units gives us numerical measures, since the whole secret of units is that they should be countable and additively related to the total quantity that they measure. In modern times the use of units and of numerical measures as a substitute for purely qualitative comparisons derives from the work of the medieval scientist and philosopher, Nicole Oresme.

Working within the Aristotelian metaphysical system, Oresme sought a way of representing the degree to which a form was present or actualized in a thing, a measure of the 'latitude of forms'. Suppose that an iron rod has one end in the fire. How do we represent the fact that there is a hot end and a cold end and gradations of heat in between? Similarly there are problems about the representation of gradations of colour and of changes in velocity. Oresme's idea was to lay out along a base line the distance along the body in which gradations of quality occurred. Then he erected perpendiculars on the base line whose heights represented the degree to which the form was manifested in the body at that distance. In this way gradations of qualities like warmth or colour, or what we should now regard as quantities like velocity, could be readily represented. In fact this was the beginning of the graphical method. We can but conjecture about the actual influence of Oresme upon the development of quantitative ideas. Suffice it to say that from his time on, that is, from the beginning of the thirteenth century, increasing efforts were made in the direction of numerical measurement.

The essence of the method is that there should be a repetitive use of a standard unit. Increase in warmth should be treated as so many equal increments of degree of heat, so that an increase in warmth might be described as being an increment of 47 degrees, that is an increase by 47 of the equal units of increment of degree of heat. There is a twofold problem to be solved before this method will work. (i) How is a unit to be defined? (ii) How is the unit, once defined, to be maintained, in short how is a standard once created to be maintained?

In the first place standards are *laid down*. There is no such process as the discovery of a standard. Every standard is related to something which is *chosen* or picked out for the purpose. If it can be found whole in nature, so much the better, but it is still a matter of choice that such and such a thing or process should be the standard. A natural and an artificial standard differ only in that the one is chosen from nature and the other is manufactured. The standard of time is a natural standard in that it is the natural process of the motion of the earth around the sun that is chosen as a basis. The standard of length was an artificial standard in that it was a length that was actually constructed out of metal and then said to represent the standard. Nowadays the standard of length resembles the standard of time in being based upon an arbitrarily chosen natural phenomenon. 'The metre is now defined as 1,650,763.73 wavelengths of the orange-red line of krypton-86.'[1]

The relation of the standard to the natural process upon which it is based may be quite complicated. It is a fact that 'the standard unit of time for the physical sciences is the mean solar day',[2] but the mean solar day is a mathematical abstraction from the actual motions of the earth and the sun.

The measurement of length is simplified by the fact that when one is using some sub-standard such as a metre rule the very same rule can be moved around the site of the measurement, and indeed the measuring could in theory be done with one metre rule alone, provided some rule of sliding and turning and invariance of length under that treatment could be assured.

[1] *Van Nostrand's Scientific Encyclopaedia* (Princeton, N.J.: Princeton University Press, 1968), p. 1912.
[2] Ibid., p. 1857.

But can the same be done with time? Suppose we settle on a standard second in the year 1965. Deciding that it is to be a certain fraction of the mean solar day we proceed to build clocks that divide that quantity for the year 1965 into this fraction. Next year we use the clocks again, and now *they* define the standard second. But how can we be sure that they have not changed their rate? The seconds of 1965 have gone for ever and it is logically impossible to have one of the seconds of 1965 with which to check the seconds of 1966. One could imagine a civilization in which it was held that the days and nights were of an equal length throughout the year, but that in the summer when it was hotter the clocks ran faster during the day seeming to cover more hours, and in the winter they ran more slowly due to the cold, thus seeming to leave fewer hours in the day. And in a similar way all natural processes were speeded up in the heat of summer and slowed down in the cold of winter. It seems to me that this account, when properly elaborated, could hardly be faulted. It would involve some new laws of nature, but so far as I can see they could be self-consistent.

Our faith in the steadiness of clocks and other repetitive processes is truly a matter of *faith*. Of course we back up this faith by our observations that the clocks are not being interfered with and we try to preserve our clocks in conditions in which they will not be upset by external influences, such as dust or stray magnetic fields. But none of this guarantees that the clocks, of themselves do not run either slow or fast or indeed erratically. Deeply embedded in our science is a principle, a belief of cosmic proportions and of the greatest significance, and it comes to the rescue here. It is the belief that unless a thing, material, or process is interfered with, it will continue to run unchanged. We believe that things and processes do not change of themselves. This needs some modification because we also recognize as very fundamental the category of agency, and we have recognized this in our discussion of the metaphysics of causality in the last chapter. The generative conception of causality and its associated notion of power are clearly counter to the idea that all things are always stable in the absence of external influence. We choose to make our clocks out of non-spontaneous materials. It would be crazy to construct a clock out of radioactive metals that we knew would spontaneously change their character with time, and it would be equally silly to try to construct a clock out of explosives. We choose the most inert metals and jewels, that is the strongest and the least liable to fatigue, with which to build it. It is a choice, and ultimately an arbitrary choice at that, by which we devise our standards, but even in the case of the most intractable standards of all, the standards of time, we take care that our choice is rational. Finally, as science progresses it is possible to see that certain choices of standards serve to simplify the laws of nature while others would complicate them. An excellent account of the way this works can be found in *The Philosophy of Space and Time* by H. Reichenbach.[1]

Units can be set out within continuous magnitudes. Temperature, length, time, and so on are all continuous magnitudes, and the possibility of numerical measures depends upon the division of the magnitude into countable steps by the use and repeated application of the unit. One way in which the Corpuscularian philosophy has combined with the need for numerical expression is in the idea of discreteness, the basis of the justification of the method of counting individuals by numbering them one by one. In this way comparisons of the size of sets of individuals can be made by counting the members, whose individual differences are not considered in this process. This method can be used both with real individuals, such as the members of populations, or it can be used in less obvious cases by creating a new sort of individual by laying down boundaries which do not exist in nature. For instance, when Mendel was doing his experiments with peas and counting the number of individuals which were green and smooth and comparing this number with those that were yellow and wrinkled, the count was made possible only because the discrete character of yellowness, manifestations of which were being counted, was created by ignoring the great differences in yellowness which are always present in any natural collection of things. Somewhere the line has to be drawn between yellow and green, so that a count can be made. So even with the counting of individuals there are arbitrary layings down of standards by which an individual is recognized.

Here we can speak of the difference between qualitative and numerical identity. Two peas are qualitatively identical if they are indistinguishable in colour, size, shape, texture, and weight,

[1] H. Reichenbach, *The Philosophy of Space and Time* (New York: Dover, 1958), ch. 1, § 4.

according to the criteria of difference that have been laid down. It is obvious that things which are qualitatively identical may be numerically diverse, that is there may be two or more individual things exactly alike. When we count, for scientific purposes, it is usually the number of numerically distinct individuals which are qualitatively identical, in some particular way, because, of course, no two individuals which we can find in nature and can study empirically are exactly alike. There is one class of individuals of which this is not true. It is a basic principle of subatomic physics that all electrons are exactly alike, that is, they differ only numerically. No electron can be tagged or marked in any way which would enable it to be marked off as different from its fellows. Our practical inability to tag electrons is very likely a consequence of our ignorance of their nature, and there is no reason to suppose that were we able to study electrons closely, should they turn out to be individual things, they would not show identifiable characteristics that marked them off as individuals. This is an instructive example for a more general point. Physics is a science particularly prone to elevating some fact, which is taken to be very important by physicists, into the status of a principle, when it seems quite immune from revision. No principle is absolutely immune from revision, nor has any principle an absolute necessity. When the progress of physics displaces electrons from their present status as ultimate entities, it will no doubt be possible to individuate them should they prove to be things. Even the principle that two things cannot be in the same place at the same time may be abandoned under certain circumstances.

Having identified our individuals and set about counting them there are still further assumptions involved in expressing the results of our counting as a definite numerical result, as a truth about the composition of a class of qualitatively similar things, of whatever kind it is we are counting. Real things come into existence and decay and disappear. How do we ensure their stability for the purposes of determining a sum? It is no good offering a number as the sum of a collection when we are unsure that the individuals we began with in the accounting still exist or are still countable as members of the class we are numbering. Should the census of the population be extended to include the churchyard? Should unborn children be included in the count of the population? If not why not? There is no reason we can give that would be conclusive, no special characteristic which the unborn and the dead have which precludes them from being counted as part of the population. What we do is to lay down by fiat, by decree as it were, what is to count as a person for the purpose of the census, and this is itself determined by the purposes to be served by the census. If the census is to be instrumental in determining government policy as to food and water requirements or as to housing, then it is not necessary to include the dead. If it is for the purpose of calculating the number of persons of military age, and of school-leaving age in the next three years, then there is no point in including the unborn. But if, as among the Mormons, the purpose is to baptize everyone who has ever lived into the Mormon faith, then the census must include the dead. The purpose for which we are doing something, determines the form of many of our activities. And nowhere is this more strikingly exemplified than in deciding what is to count as an individual for the purposes of some numeration. Thus stability is not ensured by finding specially stable individuals to count but by deciding beforehand what criteria will have to be satisfied for us to say that an individual has come into being, or that it has changed so much as to cease to be one of that kind.

Having ensured stability by the adoption of a convention, how do we ensure uniqueness? Having ensured that the ones we have already counted do not cease to exist as things of that kind, or indeed as things at all, how do we ensure that we do not count the same thing twice? Having taken away individual differences to make our individuals qualitatively identical as members of their class how can we give it back in just that measure that is required to make sure that each individual is counted once and once only? We can take both practical and metaphysical precautions. Metaphysical precautions depend upon the important metaphysical principle which partly determines our concept of 'individual thing', that no two things can be in the same place at the same time. Hence counting everything *in its place* is a metaphysical precaution. Another way of ensuring uniqueness which also depends upon the principle is that the counting should be done at one place only, with suitable receptacles from which the things which have already been counted could not escape. Making receptacles escape-proof would be part of the practical precautions, but the point of using them depends upon the metaphysics of things and their places. If a count could be performed in one single act then everything would be at some place, and not in the same place as any other thing, and the count would have to be of unique things. But counting takes time, hence the need for the practical precautions of escape-proof receptacles and valves, or lock gates, which control entrance and exit to the counting area and the receptacles allowing it to take place only one way. But such devices as the two pens and the gate between with which the shepherd counts his flock work only if another metaphysical precaution is observed. The shepherd's method works only because no one thing can be in two places at once. This metaphysical principle which is also part determinative of our concept of thing ensures that such practical methods as that of the shepherd actually work. Is this because it is a fact that no one thing can be in two places at once, a fact that might have exceptions? Might it turn out that somewhere a thing can be in two places at once? Not at all, this is a conceptual matter, a metaphysical matter. It is a reflection of how we use the concept of an individual thing. Were there to be two qualitatively identical individuals at different places at the same time we are entitled to say that there are two numerically distinct things. While we subscribe to our conceptual system we could not treat this fact as refuting the principle. The principle cannot be refuted because it is a metaphysical principle and expresses the way we use the concepts of thing and place.

In discussing the problem of settling upon an adequate standard of qualitative identity so that the individuals of a class or set can be recognized and so be capable of being counted, we touched on the problem of borderline cases. How do we decide whether a yellowish-green pea is yellow or green, for the purposes of the count? The obvious way of dealing with characteristics which shade into each other is to choose a standard example of the characteristic A and a standard example of the characteristic B, so that B excludes A, and A excludes B, and then proceed by degree of resemblance. So, if an intermediate case is more like A than it is like B it is classified as an A and counted as one of the As for the purposes of making a count, say of the proportions of A to B in a population. Then if we suppose that some cases roughly intermediate between A and B can be divided so that each is assigned arbitrarily to A or to B so as to eliminate intermediate cases, a count can always be made.

But it is not always as simple as this. There is the difficult case of what has come to be called the 'inconsistent triad'. This case arises when we have three objects, call them A, B, and C, which when examined together are such that we find that A matches B in some particular, say colour, and B matches C in the same particular. But instead of finding, as we would normally expect, that A would then match C, we find that they do not match in colour. Such cases are not uncommon, and have come to be called inconsistent triads. They make it impossible arbitrarily to assign the intermediate case B to either A or C, or so it seems.

To resolve this problem we must ask what sort of explanation we would give of the existence of an inconsistent triad. Why does it happen? The obvious, and probably the right, answer is that our powers of discrimination of, say, wavelength as colour are only of a certain fineness. We can discriminate a wavelength difference of the order of that between A and C as a colour difference, and though there is a wavelength difference between A and B, and between B and C, the difference in the wavelength of the light is not such as to be discriminated by us as a difference in colour. The same effect can be obtained with the differences in pitch between sounds, and even in the space between points as discriminated by touch. If we have no other guide in our classifying than how the inconsistent triad *appears* to us, then following out the idea behind this sort of explanation, we can proceed to resolve the anomaly by arbitrarily assigning B to either A or C. Then we say that B is really intermediate between A and C, but that the differences are not discriminable by our senses. However, we know that it is really an intermediate case because of the existence of the inconsistent triad.

The Corpuscularian philosophy is also instrumental in directing scientists towards a particular method of dealing with order scales of quality. We have noticed how we are driven towards a general replacement of qualitative by quantitative means of expressing the differences and distinctions between various states of things. We have seen this in the use of units, in the imposition of discreteness, and now we shall see this idea at work in the way measures of qualities are set up by the use of analogy.

The measurement of temperature is a particularly instructive example. Things do feel hotter or colder at different times, and different things are perceptibly different in felt warmth at the same time. There is a famous old experiment which shows how unsatisfactory it is to rely on our feelings of warmth and cold. Take three bowls, put hot water in one, cold water in the other, and warm water in a third. Keep a hand in each of the extreme bowls for a moment and then place both hands in the intermediate bowl. The water in it will feel cold to the hand that has been in the hot water, and hot to the hand that has been in the cold water. Thus our feelings give contradictory indications as to the actual state of the water. These difficulties can be resolved by the use of some process which parallels increase in temperature by some other process which is not as anomalous as the feeling of heat. One is to hand, indeed there are several processes which change with temperature, that is with degree of heat. For example electrical conductivity changes in this way, and even more useful for simple purposes, the length of a column of liquid or of gas also changes in a regular way with temperature. Now the length of the column is an analogue of the degree of heat. It serves as a measure of that degree. And, what is more, it is a lengthy measure, having satisfactory units capable of dividing a continuous magnitude in an unambiguous way and finally being simply additive; so that temperature increase can be marked by simple addition of units of temperature. The standard can be defined arbitrarily in several convenient ways, as, for example, the amount the column expands between $4°$ and $5°C$.

This device or instrument for measuring temperature leads to a change in the concept of temperature where a purely qualitative concept is changed into a quantitative one in two main stages. In the first stage the old concept of temperature is modified by reference to the analogous process and scale, so that anomalies are removed in favour of what the temperature of the middle bowl *really is*, that is what grade is reached by the expansion of the liquid or gas in the column of the thermometer. But another stage now comes to pass, when a theory is introduced to explain how it is possible for an expanding column of liquid or gas to increase in length with degree of heat. The theory introduces some new, theoretical concept of temperature. In this example it is a concept having to do with the internal energy of the material in whose temperature we are interested. This concept explains not only the laws which describe the changes in degree of heat among bodies but also explains the thermometer. It explains why the expansion of a column of liquid or of gas is an analogous process to the increase in degree of heat, by showing how it is that increase in degree of heat is the cause of increase in the length of the column, the two being related in such a way that the one can be a measure of the other. Finally if it is a good theory, like all good theories, it must explain the anomalies and exceptions to which the correlation is subject, not only that of the paradox of the three bowls, but also such facts as that water increases in density, that is contracts in volume between 0° and 4° C, and thereafter decreases in density, that is expands.

The differentiation of instruments

The thermometer is one among thousands of instruments which are in daily use in science laboratories for the measurement of as many features of the systems under study as can be treated according to the principles I have set out, and which are thereby capable of numerical expression, in short of such characteristics as can be measured. From a philosophical point of view instruments are of two main types. There are those which I shall call 'self-measurers' and there are 'non-self-measurers'. The reasons for this choice of name will become clear.

Self-measurers

In measuring length by means of a rule or metre rod we use a length to measure a length, so the metre rod is a self-measurer. The most primitive instrument is a single metre stick or rod which is laid time after time along a space according to a rule of movement. The length of the object is the number of times that the stick has been laid along it. The rules are designed to ensure that the measuring is all done in one way, so as to avoid anomalies. The simplest rule is that the measuring should start at one end of an object, then proceed by sliding the stick along the line of its own length until its beginning coincides with a point which marks the former place of its end. But other rules could be devised, which would give different results but avoid anomaly. We could still lay a carpet correctly in a room provided both were measured according to a rule which demanded that every second sliding should be accompanied by a turning of the rod clockwise through a right angle to the direction of sliding, and by rotating it anti-clockwise, turn it back into the original direction of sliding. Not only are there rules for the use of measuring-sticks but there are also metaphysical and physical assumptions involved in their use. One such assumption is that the stick does not change in length during its transport over the body that is being measured. This is a rather blunt assumption, and we could assume the rather weaker principle that if there are changes then they affect all objects equally, so that if towards one end there is a force stretching things out it stretches the stick and the object to be measured in just the same way. This would allow the use of a rubber metre stick to measure rubber blocks if they were both to be put under stress together. But at the heart of all practical methods is the assumption which usually makes measurement possible: that moving a measurer about in space does not alter its length.

The graduated ruler and the marked measuring-tape are more common instruments for measurement than is the metre stick. They are much more complicated logically. In essence they combine two different devices. Each is a measuring-stick, and each graduation of the tape or of the ruler is equal to the others and is equivalent to the sliding of the original unit into a new position, where its beginning coincides with the place where its end formerly lay. But in addition to the graduations each of these instruments numbers the marks. The numbers allow the total number of units to be read off immediately. They are nothing less than a simple computer which automatically tallies up the number of slidings forward which would have had to be made by the measurer in bringing his standard inch or centimetre forward until at last its end coincided or nearly coincided with the far end of the object to be measured.

The only other kind of self-measurer in common use is the clock and that is logically speaking also a very complex device. At the heart of a clock is the escapement, which by going 'tick, tock' creates a sequence of events which is the lapse of time. The rest of the mechanism introduces an analogue to the passage of time, namely a translation in space, by using the escapement to ensure that there is a uniformly moving point which traces out equal spaces in equal times.

The clock is made into a computer by adding numbers around the face so that the number of seconds, minutes, and hours that elapse is automatically totted up. An ordinary clock is a self-measurer, an analogue device, *and* a computer.

Non-self-measurers

A thermometer is a non-self-measurer, because a temperature or degree of heat is not being measured by a degree of heat, it is being measured by a length. And the scale adds the computer. Such an instrument works by using a length or time as an effect, the cause of which is what is ultimately being measured. A dynamometer measures the power output of an engine by computing one of its effects, namely the work that it can do in a unit time. Similarly a kilowatt-hour meter, a familiar domestic device, works by computing the total of a certain time effect, the rate of rotation of a disc, which is the effect of the rate of consumption of power, which we are ultimately concerned to measure. Behind our confidence in such instruments lie not only all the assumptions we have noticed about the class of self-measuring instruments, since a self-measurer of length or of time is the final end of the process, but there are also assumed all the theories which are our basis of confidence in the use of the effect as a measure or indication of the cause. Unless we believed that the rate of rotation of the disc in the kilowatt-hour meter was an effect of the rate of total electrical flow we could hardly have any confidence in the capacity of our meter to register it.

Do our meters give us objective information about the world, information which is independent of theory? Of course they do not. There is no such thing as information which is independent of theory. Of course we have information from our meters in which we have confidence, but that is because our knowledge bears upon such a phenomenon as the flow of electric current, from a great many different points of view. Many instruments and many theories converge upon the phenomenon we wish to study. And it is the conciliance of the results of our experiments and of our theoretical calculations, converging from widely different effects, that is the basis of our confidence. It is the way everything hangs together that counts. But the history of science shows that *merely* having everything hanging together is an illusory guide to truth. The perfection of conciliance must be combined with the constant intervention of instruments and their readings, which after all might be different from what we expect. It is in the fulfilment or non-fulfilment of our expectations as to the effects of the phenomena we are studying that objectivity lies, not in the measures themselves.

One of the central doctrines of the Corpuscularian philosophy is the distinction between primary and secondary qualities, that is between qualities which in our perceptions are truly the properties of things themselves, and those which are different in our perceptions from their causes in things. Bodies really do have shapes, and though we may not perceive their shapes exactly as they are, nevertheless the things we perceive do have shape, and from the shapes that we perceive we can readily enough work out what shapes things really have. The same goes for number and for motion, and for volume. But as we have seen, what is colour to our way of perceiving, is, in fact, a complex electrical property of things. The cause of our perceiving colours is not colour, or to put it less paradoxically colour in a thing is a different sort of quality from colour as we perceive it. It is electrical in things and to our eyes it appears as the hue of a surface. Heat is a motion but in perceiving warmth we perceive a different quality from motion, namely warmth. This distinction partly coincides with another, upon which the Corpuscularians also insisted, and which is the basis of the third surviving Corpuscularian idea. It is that many qualities vary with the state of the subject, the perceiver, while for scientific purposes we should choose those qualities which are subject invariant.

Galileo described an example of this which is still particularly instructive.[1] Consider, he says, a statue and compare it with a living man. Take a feather and tickle the statue and tickle the living man. On some parts of his body the man does not feel a tickling sensation, but there is the mechanical interaction of one material thing in contact with another. So a man can be aware of mechanical contact, the same kind of contact that he sees between the feather and the statue. But on other parts of the man's body the feather is felt to be tickling. The man reacts by flinching or laughing, or by some other response, indicating that he is experiencing a special sort of feeling. The statue shows no such reaction and does not feel the tickling. So the experience of the tickling by the

[1] G. Galileo, *The Assayer*, in *Discoveries and Opinions of Galileo*, ed. Stillman Drake (New York: Doubleday, 1957), p. 274.

man is in part due to the contact of the feather with his body, a
kind of contact he has in common with the statue, and in part due
to the fact that a man is a sensitive organism. So the tickling is not
an objective property of the feather but a product of the objective
properties of the feather with the sensitivity of the man. And what
is more, its degree differs in different parts of the man's body and
at different times. It is less perhaps when he has drunk a good deal
of wine. There are subject variant qualities and subject invariant
ones. The subject invariant are coincidental with those qualities
which are involved in the causal relations between all kinds of
bodies, while the subject variant ones are coincidental with those
which are involved in the causal relations between bodies, one or
both of which may be sensitive organisms.

This idea would enjoin the replacement of subject variant with
subject invariant qualities, and can be achieved only by the
simultaneous achievement of the substitution of quantitative for
qualitative descriptions, in particular the replacement of qualitative scales by quantitative analogues. Degree of warmth is a qualitative scale, subject variant, and capable of illusory and even
delusory appearances, while temperature as measured by a thermometer is a quantitative scale and subject invariant. This is because errors due to the subject, such as parallax, can be removed
by attention to the conditions under which the observation is
made, and the correct reading of the scale can be worked out from
that appearance.

All three Corpuscularian ideas interlock into one overriding demand, that the ideal form of description of the world should be
in terms of the metrical geometry of structures of standard elements and that such structures should be considered to be what
the things in the world are really like. It seems that by striving to
fulfil this joint ideal we are passing towards knowledge of the
world that is as objective and truthful as we could find. It seems
that real knowledge consists in knowledge of the measurements of
the relations in space between the ultimate corpuscles. But we
have seen this as a development of the Corpuscularian philosophy,
as shot through with a certain metaphysical theory, and finally as
penetrated throughout by assumptions and beliefs in particular
physical theories without which most of the experimental results
we obtain would not make sense.

At the heart of the reservations about objective truth which

I have just been rehearsing lies the fact of the dependence of science upon particular observers and upon particular theories. Two famous proppsals have been put forward in the past to try to eliminate these dependencies and so to make knowledge more objective. Both were theories about the meanings of words, and both hoped to establish ways of giving meaning to the words that would figure in a descriptive vocabulary for the use of scientists that would fulfil an ideal of total objectivity.

The theory of ostensive definition argued that if words could be given a meaning by direct confrontation with the world and in direct relation with things and qualities, then there would be no ambiguity in their use and no theories involved in applying them. So the idea was put forward that for the really basic vocabulary of science words should be given a meaning by a process called ostension, that is by pointing to an example of the quality or thing for which the word was to be used and so fixing its meaning objectively. 'Red' would be taught by pointing to a sample of the colour red and so would be given a meaning which was objective. It was thought that a whole vocabulary of descriptive words could be created in this way from which elementary sentences could be constructed. It was supposed that there could be no doubt of any kind as to the truth or falsity of the utterance of such sentences in any particular situation.

A little reflection on this theory shows how unsatisfactory it is. Of course pointing to samples does play a part in the learning of words, but what part exactly? It cannot be the whole part, since wherever a finger points there are many qualities, relations, individuals, and materials, any one of which might be what was sought. If we already have some ideas about the world and some conceptions of language and indeed some metaphysics, such as the categories of thing and of property or quality, then pointing can play a role by introducing a sample of what the word is to be used to describe. A word cannot be defined by pointing, but by pointing a paradigm of its use can be introduced. Once one knows that 'red' is a colour word, then pointing to a Soviet flag can fix the concept by showing what particular colour red is. But no amount of pointing can introduce the concept 'colour'. That cannot be learnt ostensively. How it is learnt I have no idea.

The other famous attempt to make knowledge more objective is the theory of operationism, with which we have already had a preliminary skirmish in Chapter 3. As I explained there this theory has been proposed at various times in the past, particularly by Sir Benjamin Brodie, the chemist, in the nineteenth century, and in the twentieth century by P. W. Bridgman, a physicist. In essence it takes its start from a consideration of experiment. What, asked Brodie and Bridgman, is the most objective feature of an experiment, which does not depend upon the kind of theories upon which, for instance, the kilowatt-hour meter depends, or the kind of assumptions that we have called metaphysical? It is, they both replied, the operations that the experimenter performs, in order to generate the set of numbers which are his results. A chemical compound. Brodie advocated, could not be objectively described by describing its mythical atomic constitution, which was quite unobservable and only known by a long train of dubious inferences involving not only general metaphysical theories but many particular scientific theories as well. An objective description could be found by describing the operations in the laboratory which led to its production. In a similar vein Bridgman advocated defining the descriptive vocabulary of physics by a description of the operations carried out to make a measurement. Theory, for both Bridgman and Brodie, played the role of a bridge between the different sets of numbers generated by the carrying out of the operations, and was not to be taken seriously as a deeper description of reality or of the causes of the phenomena to be studied.

We have already looked at some of the defects of this theory. But there are two which are of great importance in this context.

The first concerns one of the most striking features of science, even of quite simple science, namely the way in which quite different kinds of operations yield the same or similar set of numbers. This is explained in ordinary science by the assumption that when this is so, the explanation is that the same objective characteristic of the world is being measured by different methods. For example, the length of a field can be measured by the use of a tape or by the use of a theodolite, plane table, and base line. The coincidence of the results of the two different operations of measurement is explained by the fact that both are different ways of measuring the same characteristic of the field, namely its length. But for an operationist, since they are totally different sets of operations, they must define two different empirical concepts, and the close similarity of the results must be a remarkable coincidence. It would be a mystery which would be impossible to unravel, because any third set of operations which might resolve the mystery would define a third empirical concept, a third characteristic of the field, and so the mystery would deepen.

One of the great powers of theories in science is to be the basis of the design of experiments. Theories suggest which experiments are worth doing, and what to expect by way of results. But a full-blown operationism allows any set of operations to define an empirical concept, and there is no way by which these sets of operations can be graded as potentially fruitful or dismissed as silly. For the operationist, any set of operations, however random in genesis, that generates a set of numbers, must define an empirical concept, an objective feature of the world. So one way of advancing science would be to make random assemblages of apparatus collected in random ways, recording all the numbers which were obtained by running the equipment, when it would run. And theories could be constructed by simple analytical methods which would combine the different sets of numbers under some general principles of an algebraic kind. All this is quite ridiculous.

The upshot is that we must recognize that there is no purely objective knowledge, if by that is meant knowledge in which metaphysics and the actual theories of science, old and new, play no part. Even in the simplest parts of science, expressions like 'current' in electricity already carry with them the suggestion of fluids in motion, and like 'force' in mechanics, of effort expended, which are the ghosts of long dead theories. The legitimate ideals of objectivity are not to be achieved by denying the *a priori* elements in the scientific description of the world, but by recognizing them for what they are, and while realizing that they are indispensable, understanding too that when particular theories and assumptions prove to be a bar to understanding or progress they can be changed and improved. What we cannot do is describe the world in the absence of any prior understanding of it, and in the absence of any theory.

Summary of the argument

1. *The principle of structural explanation*

The properties of things and materials should be defined, as far as possible, in terms of structural relations among a small number of elementary units.

(*a*) A typical application is the definition of such a quality as heat in terms of the motion of insensible parts, of allotropy as due to differences of molecular structure of standard atoms, e.g. carbon.

(*b*) The preferred form of explanation of manifested multiplicity, is as a multiplicity of structure among the fewest possible basic kinds of entity, e.g. the explanation of the multiplicity of chemical elements.

(*c*) Adherence to the principle of structural explanation entails adherence to the principle of paucity of kinds of entities. Associated with this idea, but not entailed by it, is the principle of paucity of kinds of relations among the elements, as e.g. behind Newton's attempt to reduce gravity to action by contact.

(*d*) From (*b*) can be inferred immediately a criterion for the completeness of an explanation, i.e. when all kinds of diversity are explained as cases of structural diversity among only one kind of entity. Ideally the structure will be a space–time structure. This is the most refined form of microexplanation.

(*e*) Two different relations between the properties of the parts and the properties of the whole are possible:

(i) Additivity: the property of the whole is an additive function of the properties of the parts, as e.g. mass.

(ii) Emergence: the property of the whole is produced by properties of the parts but is not qualitatively similar, e.g. heat and motion, the power of speech, and the powers of individual human nerve cells, etc.

(A) The explanation of emergence is that the holistic properties derive from the structure, and not from the entities structured.

(B) The explanation of other emergent properties has to do with the unique ensemble of simpler properties. Living things, for instance, contain nothing but chemical reactions, but their liveliness derive from the particular and mutually supporting ensemble of these reactions, the explanation of which is not chemical.

(*f*) Real science also contains macroexplanations.

(i) In social science the behaviour of a member of a group may be explained by reference to a function that the group performs.

(ii) In physics the explanation of inertia as a function of the effect that all the masses in the universe have upon any one is a macro-explanation.

2. *In an ideal form of description, quantitative items replace qualitative items.*

(a) This is manifested in the preference for measures using units, rather than simple ordering scales.

(b) This ideal antedates the Corpuscularian philosophy, and is found in Oresme's theory of the latitude of forms.

(c) The use of units raises two problems:

(i) A practical problem of the definition of a unit must be solved.

(ii) A metaphysical problem of the justification of taking it as invariant at different parts of the scale must be solved.

(A) Solved for length by defining an abstract unit with respect to all known correction requirements.

(B) Solved for time by *a priori* choice, guided by empirical knowledge of possibly distorting influences, e.g. clocks are not made of material known to decay spontaneously.

(d) Quantitative measures can also be obtained by counting.

(i) This requires conventions as to the boundaries between types, and hence as to discreteness.

(ii) Associated with this will be conventions as to which qualitative differences are to be discounted, in identifying 'pure' numerical difference.

(iii) The use of the result of a count as a measure of the composition of a class depends upon assumptions about the permanence and identity of the entities involved. These depend upon the nature of the problem to which the count provides a solution, e.g. should the dead be included in a census? Note that the Mormon census does include them, for the purpose of retrospective baptism.

(iv) Practical precautions to prevent counting something twice, or not counting it at all, depend upon metaphysical assumptions, such as the principle that no two things can be in the same place at the same time.

(v) There remains the problem of the inconsistent triad, A, B, and C, where A matches B, B matches C, but A does not match C. This is solved in practice by the assumption that A and B, B and C *really* differ, but by a degree finer than whatever was being used to discriminate between them.

(e) This assumption lies behind the replacement of fundamentally qualitative scales, such as degree of felt warmth, by quantitative scales, such as temperature.

(i) Such a step involves the empirical fact of analogy between the expansion process and the degree of felt warmth, while conceptual stability is maintained, i.e. while the thermometer is conceived of as a measure of the quality *heat*.

(ii) Then a new concept grows up around the instrument, namely temperature, which

(A) is embedded in theory, which explains the original analogy,

(B) is used to refine and diversify the original qualitative concept, as e.g. in the aphorism, 'It's not the heat but the humidity.'

3. *The differentiation of instruments*

(i) Self-measurers: using a length to measure a length, or a process to measure time involves invariance assumptions discussed above, and a standard procedure to ensure comparability of results.

Usually a self-measurer contains a computer as well, which automatically delivers the result of an assumed sequence of operations each of which would be counted, e.g. graduated ruler, clock.

(ii) Non-self-measurers: using a process of change which is analogous to what is to be measured, grounded in the assumption of a causal relation between the one process and the measurer, e.g. thermometers, wrist-calcination as a measure of biological age, etc.

4. The philosophical background to the Corpuscularian philosophy depends upon the distinction between primary and secondary qualities, and this distinction was made by differentiating those qualities which seemed to depend upon the observer, and those which were supposed to be independent of him. This distinction has not survived criticism, but the forms of description it generated have.

(i) The ideal form of description should use subject-invariant, rather than subject-variant quality terms.

(ii) Introduced by Galileo, in modern times, with the example of the distinction between mechanical contact between things, and such products of interaction as tickling.

(iii) This ideal is connected with the second ideal, in that the choice of analogues for a process is partly determined by the wish to eliminate a notoriously subject-variant quality, as e.g. felt warmth.

5. The overriding ideal which coalesces out of the three particular ideals, is that the ideal form of description of the world is a metrical geometrical structure among standard elements, whose properties are reduced as far as possible to the mere occupancy of space. These structures are embedded in a field of potential, and modify it.

6. The ideal of subject-invariance for properties is derivative from

a deeper epistemological principle, that of total objectivity. The choice of primary qualities as the basis for science is only one way of trying to meet this demand.

(i) The theory of ostensive definition aimed at providing meaning for all descriptive terms by reference to immediate sensory experience, on the grounds that if descriptive words were given meaning only with respect to the immediate sensations experienced by their users, then there could be no room for error. We have examined the metaphysics of this view as phenomenalism. As a theory of definition it fails too, since pointing is insufficient to differentiate the category of experience to which the introduced word is supposed to refer.

(ii) The second attempt to achieve objectivity automatically by means of a theory of meaning and definition of scientific terms is operationism, either in the Brodie or the Bridgman form, which we have already examined as epistemological theories. As theories of meaning they are defective because

(A) if each set of operations defines an independent empirical concept, then different ways of measuring what in pre-operationist terms is the one physical property, yield similar sets of results, whose similarity is an insoluble mystery on operationist grounds;

(B) the design of experiments is detached from theory, since *any* assemblage of equipment and sequence of operations defines an empirical concept. Thus there remains no way of differentiating absurd from fruitful experiments, indeed that very distinction has no place.

6
Explanation

WE HAVE BEEN feeling our way, from several different directions, towards the culmination of our study, the elucidation of the ideal form for theories. Theories are the crown of science, for in them our understanding of the world is expressed. The function of theories is to explain. We have already identified some of the forms of explanation.

Two important paradigms of theory have appeared upon which we can base our ideas of what a theory should be. Here I refer back to our discussion of Chapter 3. The science of mechanics with its central concept of force is one, and the science of medicine with such concepts as the virus is another. They are opposing paradigms as we have had occasion to notice already. They present two different kinds of theory as seen from a logical, epistemological, and metaphysical point of view. Must we accept both paradigms? Does each have a particular role to play? Can the one be reduced to the other? We shall come some way to settling these questions in this chapter.

The concept of force and the concept of virus seem to play similar roles since each is used to explain observations, on the one hand concerning motion, and on the other concerning the course and development of disease in plants and animals. In the normal course of events neither a force nor a virus is observable in the way in which the happenings which they are designed to explain are observable. Finally, both the conception of force and the conception of the virus are concepts devised by analogy. They are descriptive of entities analogous to certain things with which we are familiar. Forces are analogous to the efforts that people make in shifting things against a resistance, and viruses are analogous to the bacteria which had been found to be the causes of many diseases. We have looked already at some of the detail in the development of these analogies. But if we look a little further at the science of mechanics, and compare it with pathology, a deep difference appears. The concept 'force', and with it the analogy with human effort, is inessential to the science of mechanics, as has been shown by the several ways in which that science can be reformulated without this concept. Its function is entirely 'pragmatic'. It serves the function of an aid to understanding, a device by which intuition is engaged in the business of understanding motion. But it is perfectly possible to understand motion without the concept of force. It is possible to understand all the phenomena of motion using, say, the concept of 'energy', and its redistribution among the bodies involved in a system of moving particles according to certain laws. The analogies are quite inessential to mechanics. But compare this situation with that in pathology. Without the concept of the virus as micro-organism the whole theory of the transmission and cause of a wide range of diseases would be quite different. The theory is an essential part of the understanding of the observations. A description of a disease is one thing, its pathology is quite another. This shows up for example in the difference between bad doctoring, in which the symptoms only are treated, and good doctoring, where a diagnosis of the cause of the disease is made and that cause is treated. Finally, and it is this which explains all the other differences between the concept of force and the concept of the virus, it makes sense to ask whether or not there are viruses, and it makes a tremendous difference to medicine which way that question is answered. But though it makes sense to *ask* whether or not there are forces, whether or not individual things exert efforts as people do in bringing about motion, it makes not the slightest difference to the science of mechanics whether there *are* or *are not* forces. The science would be differently formulated no doubt, but there would be not the slightest difference in the predictions of future states of moving systems which could be made by means of it.

Each of the paradigms marks an important ingredient of theory, at least one of which must be present for a theory to be satisfactory. The science of mechanics is organized by the use of a mathematical mode of expression into a logical system, where there are certain fundamental principles and the practical laws of motion are deduced from these in a logical and rigorous way. It is obvious that there are certain great practical advantages in being able to express a theory in a logical system. A great many particular laws and even particular facts can be comprehended in a very economical way in the principles of the theory. Systematization has considerable pragmatic value. But a theory would still be a theory and would still explain the facts it did explain if its laws did not fit easily or at all into a logical system. The laws might only hang together because they were the laws of the same subject matter, that is the laws describing the behaviour of the same kind of things or materials. We have a great deal of knowledge about human behaviour for instance. But this knowledge cannot be formulated in such a way as to fit into a deductive, logical structure. Our theory of human behaviour is a rag-bag of principles united by virtue of the fact that they all concern the same subject matter, namely, the behaviour of people. We may never find a systematic formulation of these laws. We may never achieve the pragmatic advantages of system.

A scientific explanation of happenings, whether individual happenings or sequences of events, consists in describing the mechanism which produces them. Only in the most minimal sense does the science of mechanics explain any course of motion. The laws of mechanics are descriptive laws, not explanatory laws. Apart from the tenuous and rather feeble concept of 'force' there is no attempt in that science to advance any account of the mechanisms of motion, of why the laws of impact, of momentum conservation, and so on are what they are. So far as I know the only attempt that has ever been made at this is the grotesque set of explanations offered by Descartes in Book II of his *Principles*.[1] But the virus theory has exactly what is required for a scientific explanation of the course of the disease with which it is concerned. The presence of the virus explains what is described in the syndrome or course of the disease, and the more we know about the nature and behaviour of viruses, the more we know about the disease. It is the interaction between body as host and virus

---

[1] R. Descartes, *Principles of Philosophy*, Bk. II, xxxvii, xxxix, xl; in *Descartes, Philosophical Writings*, trans. E. Anscombe and P. T. Geach (London: Nelson, 1954), pp. 216–19.

as parasite that produces the symptoms of the disease and explains the course of it. The virus theory of poliomyelitis is truly a scientific explanation, where the beautifully systematized laws of mechanics are not. Of course in certain cases the mechanics of particles in motion explains other phenomena, because then the laws of mechanics serve as perfect descriptions of the causal mechanism at work. Such for example, is the often quoted example of the kinetic theory of gases, where the mechanics of the molecules of the gas sample serves as a causal mechanism which explains how samples of the gas behave under various conditions. The kinetic theory is an explanation, and a scientific explanation at that, of the behaviour of gases, but it follows the paradigm of the virus explanation of poliomyelitis, and not the paradigm of the force formulation of mechanics. The fact that mathematical means of expression are used in the kinetic theory and in mechanics, and are not used in the virus theory should not blind us, as philosophers of science, to the essential difference of the former and the essential likeness of the latter.

The generation of the concept at the heart of a theory, what Whewell called the Idea of the theory is, as we have seen in the many examples that were discussed in Chapter 3, a matter of analogy. Building a theory is a matter of developing an appropriate concept by analogy. This is the essential heart of science, because it is the basis of explanation. Why is it that we cannot just go out and find out what the basic mechanisms are? Why can we not eliminate the need for analogy, and go directly to nature? The answer is that science proceeds by a sort of leap-frogging process of discovery. As soon as a field of phenomena is identified as worth studying and comes under scrutiny we can find all sorts of regularities and patterns among phenomena, but we do not find among these phenomena their causes, nor do we find the mechanisms responsible for the patterns of behaviour we have found. Chemistry proceeds both in the study of the chemical behaviour of different substances and materials, and in the discovery of the mechanisms of these reactions. In studying the reactions we do not study the mechanism of reaction. In many, many cases a great many facts about a certain kind of phenomenon can be found out without it ever being possible to study the mechanisms of the phenomena directly. In such circumstances the necessary mechanisms have to be thought out, to be imagined, and to be the subject of hypotheses. And once they have been thought out, then we know what sort of observations would lead to their independent discovery. Sometimes a wholly different line of investigation leads indirectly to the discovery of the causal mechanisms underlying some phenomena. Such, for example, was the case with the study of radioactivity and chemistry, where the examination of the distintegration of certain rather unusual materials led to discoveries of the greatest importance about the structure of the elementary parts of materials, the chemical atoms. These discoveries were turned by Lewis and Langmuir into a theory of chemical reaction, a description of the mechanism of chemical bonding and the circumstances under which chemical change took place. When we do not know what are the mechanisms underlying the processes we are studying, then we must imagine them, and they must be plausible, reasonable, and possible mechanisms. To achieve this we proceed by the method of analogy, supposing that they are like something about which we all ready know a good deal, and upon the basis of our knowledge of which we can imagine similar mechanisms at work behind the phenomena we are investigating.

What is an analogy? An analogy is a relationship between two entities, processes, or what you will, which allows inferences to be made about one of the things, usually that about which we know least, on the basis of what we know about the other. If two things are alike in some respects we can reasonably expect them to be alike in other respects, though there may be still others in which they are unlike. In general between any two things there will be some likenesses and some unlikenesses. The art of using analogy is to balance up what we know of the likenesses against the unlikenesses between two things, and then on the basis of this balance make an inference as to what is called the neutral analogy, that about which we do not know. Suppose we compare a horse with a car. There are certain likenesses in that both are used as means of transport, both cost a certain amount to buy and to maintain. There are unlikenesses in that one is wholly an artefact, and only in the choice of breeding partners does the hand of man interfere in the production of horses. Horses are organisms, cars are machines. Cars can be repaired by replacing worn-out parts from an external source, but this technique is of limited application for the horse. Suppose we learn that a certain city uses only horse transport, but we know nothing else about their system. We can make certain inferences about the traffic density from what we know about cities which use mechanical transport on the basis of the likeness between horse and car as means of transport, and we can make other inferences about the air pollution based upon what we know of the unlikenesses between them. In this way, by the use of analogy, we penetrate our area of ignorance about a city whose transport is by horse.

In many cases in science we are operating from one term of an analogy only. Molecules are analogous to particles in motion, but we cannot examine molecules directly to see how far they are analogous. Since the molecule is an entity which we imagine as being like a particle in motion, we are free to give it just such characteristics as are required for it to fulfil its function as a possible explanatory mechanism for the behaviour of gases. The neutral analogy is just that part of what we know about particles that we do not yet transpose to our imagined thing, the molecule. The molecule is analogous to the particle not because we find it so but because we make it so. And there is another analogy which completes the theory. A swarm of molecules must be analogous to the gas, otherwise we should not be able to use the molecule concept as an explanatory device. These distinctions are not really well brought out in terms of the simple notion of analogy. From the point of view of the notion of analogy, the relation of molecules to material particles, and the relation of the laws describing their behaviour to the laws of mechanics, and the relation between a swarm of molecules confined in a vessel and a gas, are analogies. But whereas a gas sample might really be a swarm of molecules and molecules might really be material particles, the relationships are, from an epistemological point of view, quite different.

The distinctions which we are looking for can best be made by introducing a new concept, which allows us to analyse analogy relationships a good deal more carefully and finely. This is the concept of the *model*. In the technical literature of logic there are two distinct meanings to the 'model', or perhaps it might be better to say two different kinds of model. In certain formal sciences such as logic and mathematics a model for, or of a theory is a set of sentences which can be matched with the sentences in which the theory is expressed, according to some matching rule. We shall not be concerned with such formal, sentential models here. The other meaning of 'model' is that of some real or imagined thing, or process, which behaves similarly to some other thing or process, or in some other way than in its behaviour is similar to it. Such a model has been called a real or *iconic model*. It is with iconic models that we are mostly concerned in science, that is, with real or imagined things and processes which are similar to other things and processes in various ways, and whose function is to further our understanding. Toys, for example, are often iconic models, that is things which are similar to other things in some respects, and can indeed play something of their role. For example, dolls are often models of babies, that is, a doll is a thing which is like a baby, and can be treated for certain purposes as a baby. And baby-models can be used quite seriously in training mothers and midwives in baby-handling where it is inconvenient or even dangerous to employ a real baby for the purpose. A toy car is often a model car, a toy plane a model plane.

Models are used for certain definite purposes, and in the sciences these purposes are (i) logical: they enable certain inferences, which would not otherwise be possible, to be made; and (ii) epistemological: that is they express, and enable us to extend, our knowledge of the world. To sort out these purposes rationally yet another idea is needed, that is, the difference between the source of a model and the subject of a model. A doll is a model *of* a baby, and also modelled *on* a baby. Its source is the real thing, the baby, while its subject is, in this case also the baby. Its source and its subject are the same. Such models are called *homoeomorphs*. But when one is using the idea of the molecules as the basis of a model of gas, the molecule is not modelled on gas in any way at all. The molecule is modelled on something quite different, namely the solid, material particles whose laws of motion are the science of mechanics. Such a model for which the source and subject differ is called a *paramorph*.

Science employs both homoeomorphs and paramorphs, and indeed the proper use of models is the very basis of scientific thinking.[1] A theory is often nothing but the description and ex-

[1] For the full development of this idea, see R. Harré, *The Principles of Scientific Thinking* (London: Macmillan, and Chicago, Ill.: University of Chicago Press, 1970).

ploitation of some model. The kinetic theory of gases is nothing but the exploitation of the molecule model of gas, and that model is itself conceived by reference to the mechanics of material particles. We have seen how our lack of knowledge of the real mechanisms at work in nature is supplemented by our imagining something analogous to mechanisms we know, which could perhaps exist in nature and be responsible for the phenomena we observe. Such imagined mechanisms are models, modelled *on* the things and processes we know, and being models *of* the unknown processes and things which are responsible for the phenomena we are studying. This important fact leads to our having to acknowledge that a theory has a very complicated structure, one in which there are at least two major connections which are not strictly logical in the formal sense, but are relations of analogy. The gas molecule is analogous to the material particle, and the swarm of molecules is analogous to whatever a gas really is, and both these analogies are tested by the degree to which the model can replicate the behaviour of real gases. Gas molecules are only like material particles (in some versions of the theory they do not have volume, for instance), and a swarm of them is only like a gas, since even the most sophisticated molecular theories do not quite catch all the nuances of the behaviour of real gases.

But there is a further and final point of the utmost importance. Since we do not know the constituents of a gas independently of our model, we can scarcely be in a position to declare any negative analogy between the model and the gas of which it is a model. Any defects in the molecule concept can be made good so long as we can change its properties without contradiction. We can make and remake the molecule so that in swarms it behaves as near as we like to the gas. When we are considering a model of something we wish to understand we are presented with a neutral analogy and a positive analogy only. The fact that our model may be modelled on something with which it has, in addition to its positive and neutral analogy, a strong negative analogy is of no consequence, since it simply means that the model does express the concept of a new kind of entity or process, different from the one upon which it is modelled. Now as a model of some process or mechanism or material responsible for the phenomenon we are studying becomes more and more refined, a new question gradually presents itself. During the process of refinement we were concerned only with so adjusting our model that it behaved in a way which *would* account for the phenomena. Gradually we are brought to consider the question as to the reality of what had previously been only a model of the real mechanism of nature. Perhaps, we might say to ourselves, gas molecules are not just models of the unknown mechanism of the behaviour of gases, perhaps there really are gas molecules, and perhaps gases really are nothing but swarms of these things.

I want to present some examples now to show the different ways in which this deeply penetrating question of the reality of an iconic model can be pursued. Darwin's Theory of Natural Selection provides an excellent example of the use of iconic model building to devise a hypothetical mechanism to account for the facts which were known to naturalists. Students of nature had come to see that the populations of animals and plants that at present existed on the earth were different from those that had existed previously. They had also come to see that in nature there was a great variety of forms of many plants and animals closely similar to each other. Many people were very familiar with the possibilities of breeding, particularly gardeners, and stock-breeders of various animals. How are we to explain the variety of species that had existed, and to explain the distribution of that variety of species which now exists? What process in nature is responsible for these striking facts? Now whatever process it is works very slowly, so slowly that Aristotle, one of the greatest biologists of all time, had been deceived into thinking that the species of animals and plants were fixed, so impressed had he been by the similarity between parents and offspring. But there are also minute differences, and it was upon these that Darwin's theory was fixed. Darwin did not know what were the processes by which change in the animals and plants of nature came about, so he constructed a model. He knew very well that there is change in domestic animals and plants and he knew that that change is due to the fact that the breeder *selects* those plants and animals from which he wishes to breed, which are more suited to whatever purpose he has in mind, and that after several repetitions of selection a quite different-appearing creature can be derived from appropriately chosen individuals solely by breeding. There is a variation in nature, and Darwin conceived of a process analogous to domestic selection which could be a model of whatever process was really taking place in nature. He called this process, modelled on domestic selection, *natural selection*.

Had Darwin proceeded with the same model source, for filling out the details of his imaginary process of natural selection, he might have posited the active intervention of a breeder, who like the gardener or stock-breeder had some purpose in mind in bringing together those particular plants and animals which did breed, and which produce the subsequent generation. Now Darwin was looking for a process which was wholly natural and which did not involve divine intervention as a part of the model. He found another source which contributed to his model process. This was the theory of population pressure, originated by Malthus, and elaborated by Herbert Spencer, who had coined the term 'Survival of the Fittest', as a brief description of the outcome of the competition for space, light, food, and so on, which by analogy with the conditions which Malthus reckoned to detect in human society, could be projected on to nature. Darwin's Theory of Natural Selection became in effect the elaboration of the various ways in which different varieties of animals and plants were caused to breed at different rates, and so to explain why, in each generation certain individuals were favoured and bred more freely than others. All this was an elaboration of a basic model of whatever process was really responsible for change and development of animal and plant forms in nature. Did the things that Darwin's model suggested really happen in nature? By this question I do not mean 'Did evolution occur?' Rather I mean, 'Is it the case that some individuals are able to breed more freely because they are more suited to their environment, and do they therefore transmit to their offspring whatever characteristics favoured them?' The reality of the mechanisms of evolution as proposed by Darwin is a separate question from the reality of the evolution process, that is the gradual change of species. Nowadays it is hardly conceived by most biologists that Darwin's theory began as a model of the real processes in nature, so much is it taken for granted that Darwin's model is real. I suppose that it is still just possible, though extremely unlikely, that it may eventually turn out that quite different mechanisms are responsible for the evolution of species.

Recently an interesting example of model building has taken place in the theory of electrical conduction. Somehow the electrons that are in metals are responsible for the conduction of electricity in metal. Drude produced a very successful model of the mechanism of conduction by supposing that there were free electrons in the metal which behaved like the swarm of molecules which we have seen as a most successful model of gas. From supposing that the electrons were like a gas confined within a container, he was able, with very few supplementary assumptions, to work out an explanation of the known laws of conduction, that is he showed that a swarm of electrons obeying the gas laws would behave analogously to a conductor. Here the model is modelled on another model, and is a model of a truly unknown mechanism, the unknown mechanism of the conduction of electricity in metals.

To explain a phenomenon, to explain some pattern of happenings, we must be able to describe the causal mechanism which is responsible for it. To explain the catalytic action of platinum we must not only know in which cases platinum does catalyse a chemical reaction, but what the mechanism of catalysis is. To explain the fact of catalysis we need to know or to be able to imagine a plausible mechanism for the action of catalysis. Ideally a theory should describe what really is responsible for whatever process we are trying to understand. But this ideal can rarely be fulfilled. In practice it becomes this: ideally a theory should contain the description of a plausible iconic model, modelled on some thing, material, or process which is already well understood, as a model of the unknown mechanism, capable of standing in for it in all situations.

Finally this ideal of explanation is complemented by another and final demand. The ideal model will be one which not only allows us to reason by the complex structure of double analogy which I have described, but is one which might be conceived to be a hypothetical mechanism which might really be responsible for the phenomena to be explained. This is what prompts that deepest of all scientific questions, 'What is there really in the world? Are those hypothetical mechanisms which we believe might exist really there?'

If knowledge is pursued according to this method it will tend to be stratified. Perhaps this can be seen best if we look at the way causes are elaborated. There are two conditions which have to be fulfilled for there to be truly said to be a causal relation among happenings or phenomena. The first condition, ensuring that there is prima facie evidence, is that there should seem to be some pattern or structure in what we observe to be happening. This might be that simple kind of pattern which we call regularity or repetition, when we find one sort of happening followed regularly by happenings of a certain, definite other kind, when for instance those who are deprived of fresh fruit and vegetables develop scurvy, and those who have plenty of the above commodities do not. We have prima facie evidence that there is a causal relation between the deprivation and the disease. But to eliminate all possibility that something else, some third factor, might be responsible both for the shortage of vegetables and for the scurvy, we must find out what is the mechanism involved, and that involves us in a study of the chemistry of the food materials and of the physiology and chemistry of the body. That study supplies an idea of the mechanism which explains the pattern of happenings involving presence and absence of fresh vegetables, and the onset and cure of scurvy. Satisfying this second condition, that is, describing the causal mechanisms, completes one causal study. Our knowledge falls out into two strata as it were: in one stratum the facts to be explained are set out and their pattern described; in the underlying stratum we may imagine or describe the causal mechanism.

Now, that mechanism is described in terms of chemical reactions and physiological mechanisms. These exhibit their own characteristic patterns and regularities, and these call again for causal explanation. But now a new kind of fact must be adduced. Chemical reactions are explained by the theory of atoms and molecules and chemical valency. By means of this model we can describe a causal mechanism for chemical reactions, and similar considerations apply to the explanation of the physiological and biochemical facts. We have reached another stratum. Then that stratum itself becomes the occasion for prima facie hypotheses that there too are causal relations, that there is some mechanism which explains the combining powers of chemical atoms, and some model of the chemical atom which would explain the diversity of chemical elements. Such a model is to hand in the electron–proton–neutron picture of the atom and the electronic theory of valency. This forms another stratum. Finally, and this is where we are today, if we are to be true to our scientific ideals, we must ask what is responsible for the behaviour of protons, neutrons, electrons, and the other subatomic particles, and we must try to penetrate to yet another stratum. As we have seen in this chapter, this must first be a work of the disciplined imagination, working according to the principles of model building, the method which has enabled us to proceed to such depths in uncovering the strata of the mechanisms and processes which make up the natural world.

In each era scientists find themselves at a loss, incapable of proceeding deeper into nature. And in each era scientists explain this temporary ending of scientific penetration by a metaphysical theory in which what is basic for one time and one limited scientific culture is elevated to the status of the ultimate. As we saw in the chapter on metaphysics, the metaphysical theories of the past have presented forms of explanation as ideals, and those ideals, expressing the ultimately conceivable models for that culture, end with a seemingly impenetrable stratum, that closes the layers of knowledge. But we have also seen that metaphysical systems are not systems of facts. They are systems of concepts which we invent, and which we adopt if we will. Without them we could not think at all, but we must not allow any particular one to stand in the way of scientific progress. Perhaps science may come to an end for us, by reaching a stratum beyond which we have neither the imagination nor the technical resources to penetrate. But that end will not be the end of nature, it will be the projection upon nature of our own limitations. In the meantime we have no alternative but to follow the methods of science as we know them.

Summary of the argument

1. Since the function of theories is to explain, an examination of the structure of acceptable theories will yield the forms of explanation.

(a) Two paradigms of theory:

(i) The mechanical theory of force introduces an entity which is not observable, which is supposed to cause the mechanical phenomena, *but which can be eliminated from mechanics*, without radically changing the theory. The function of the 'force' concept is easily seen to be 'pragmatic', serving only to enlist intuition in the understanding of certain abstract relations.

(ii) The virus theory of disease introduces an entity which was unobservable when first introduced, and which is supposed to cause the observed phenomena, but which cannot be eliminated from the theory without entailing a radically different conception of illness, cure, and so on.

(b) Consequential differences between the paradigms:

(i) When we understand mechanics we come to see that the question of whether there are or are not forces does not arise, but when we understand the virus theory of disease we do so by accepting the putative existence of viruses, and commit ourselves to the hazard of having to abandon the theory if it is shown by later developments that there are no such things.

(ii) Mechanics is capable of a high degree of organization according to the principles of mathematics and deductive logic. The virus theory is united around the central entity 'virus' whose contingent features are not deducible from some set of first principles. For such sciences as social psychology there is not even, in principle, a way in which the general features of human behaviour could be united into a system with only logical connections.

(iii) Reflection on the virus example leads one to say that a scientific explanation is characterized by the fact that it describes the causal mechanism which produces the phenomena.

(c) The origin of the concept at the heart of a theory:

(i) In general the hypothetical entities which constitute the causal mechanisms referred to in scientific explanations are not discovered initially by observation.

(ii) They are first imagined, and their attributes are derived by analogy with entities already known, either by observation, or as the hypothetical entities enshrined in another explanation.

2. (a) Analogies have the following structure:

(i) Positive analogy, that in which A and B are *alike*.

(ii) Negative analogy, that in which A and B are *unlike*.

(iii) Neutral analogy, those attributes or either A or B about which we have no information as to their being matched in the analogue.

(iv) In conceiving hypothetical entities we can examine only one of the entities entering into the analogy, namely that from which the analogy derives, i.e. its source.

(v) The behaviour of the hypothetical entity must be analogous to the behaviour of the real thing which is really causing the phenomena under study,

(b) The technical concept of 'model' allows a ready codification of these conditions. In science we are concerned with 'iconic' models, that is analogues of things and processes.

(c) The two main uses of models in science are

(i) Heuristic, to simplify a phenomenon, or to make it more readily handlable, as e.g. hydraulic models of electrical networks.

(ii) Explanatory, as described above where the model is a model of the real causal mechanism, then unknown.

(d) Models can be classified by whether their source and subject are identical (*homoeomorphs*) or different (*paramorphs*). In explanatory theories source and subject must be different, so these theories use paramorphic models.

(e) Creative use of paramorphs involves no negative analogy, since disanalogies simply disappear from the definition of the model.

(f) Successful use of an iconic model begins to prompt 'reality' questions:

(i) It may be supposed that the iconic model is a true or good representation of the real causal mechanism.

(ii) Provided that it depicts a plausible hypothetical mechanism, it may be possible to inaugurate a search for the entities involved. This, in itself, may stimulate the invention both of sense-extending instruments, and of new forms of detector. As an example of the former consider the electron microscope, and of the latter, the Geiger counter.

3. This conception of a model as paramorph can be used in the analysis of theories:

(a) Darwin's evolutionary theory can be looked upon as the description of an iconic model of the unknown processes of evolution through

(A) the analogy between natural selection and artificial selection as explanations of natural variation and domestic variation respectively.

(B) the analogy between plants and animals in the world and the competition for resources among an ever-expanding human population.

(b) Drude's explanation of the relations between electrical and thermal conductivity depends on the invention of a model of the causal mechanism of these phenomena derived through an analogy between the electrons in a metal and the molecules in a gas, that latter itself the product of a famous analogy between the particles of a gas and ordinary material things in motion.

4. Organized in this way, knowledge is stratified.

(a) In the stratum of observation non-random patterns are discovered which call for explanation.

(b) Their explanation is provided by the description of causal mechanisms, in general unobservable, whose behaviour generates the observed pattern.

(c) This process of stratification continues until the most fundamental relations recognized in each era are reached.

7
Science and Society

THE DISCUSSION of science which has occupied the last six chapters of this book has been based on the presumption that science can be fully understood as an intellectual and practical activity without reference to social forces and influences of any kind. However, science is a social activity. It is carried on *by* groups of people *for* groups of people. Its results are used by communities. In this chapter I turn to a discussion of some of the philosophical problems that arise when science is looked at in this light. Not only will this help us to see certain moral and ethical issues that are obscured in our previous analyses, but it will help to make clear the difference between philosophical problems (and how we tackle them) and problems that are properly the concern of sociology and history.

The first distinction that we need to draw is between science *in* a society—and the philosophical problems that this relationship leads to—and science *as* a society. The latter will lead us to philosophical issues that appear when we think of scientists as a community, perhaps even as a tribe with its own distinctive culture and customs. First we turn to some issues raised by considering science *in* society.

Science in society

What sort or argument could be used to justify a demand that society should support science? Why should the government contribute funds to research councils, universities, polytechnics, and other places where science is carried on? To investigate this question more closely we must distinguish between justifying an activity by reference to its consequences and justifying it by reference to its intrinsic worth. For instance, our attitude to science might be based on an assessment of the improvements in public health that would come about as a consequence of applying scientific discoveries in everyday life. But we might think science was somehow good in itself. There are those who would assess the value of scientific research through an appreciation of the 'beauty' of certain scientific theories.

Science as an activity justified by the value of its applications

Looking at science in this way we see it as valuable because it is instrumental to the achievement of further goals which themselves have some kind of intrinsic value. For example, Francis Bacon thought that scientific research ought to be supported because it would produce useful discoveries that would improve the health, wealth and general welfare of the nation. It was taken for granted that such improvements were valuable in themselves. It was better that people should be healthy than diseased. If asked to justify that assessment one could perhaps fall back on some ultimate human good, such as the promotion of happiness, and the elimination of suffering. In this kind of argument for science the moral and political issues concerning the value of scientific research are shifted away from that research itself to its associated technology, since it is in its applications that science shows its human worth. Many arguments in the popular press for or against some scientific project are built up in this way. The justification or condemnation of scientific research projects is based on the moral qualities of the results of applying its associated technology. Scientists themselves sometimes argue that scientific studies are morally neutral and that ethical and political issues arise only when scientific discoveries are put to work by governments or industries. By arguing in this way scientists exempt their scientific activities from moral criticism and feel free to carry on with programmes of research even when it seems clear that the results of their work will be likely to be applied in morally obnoxious developments. Thus, for example, it could be argued that the study of mutant strains of virulent diseases is just microbiology. It becomes germ warfare only when these studies are taken up by governments and their products conceived of as weapons. But those who take up this line of argument to rebuff condemnations must be prepared to accept that consistency demands that they do not

attempt to justify their work by reference to any of its good consequences.

Science as an activity justified by its intrinsic worth

Some people certainly believe that advances in our knowledge of
nature are intrinsically good, in no need of utilitarian justifications.
The French biologist, Jacques Monod, in an influential popular
work, *Chance and necessity*, argued that scientific knowledge was a
basic moral good, so that the disinterested pursuit of knowledge was
a matter of unconditional moral worth. However, there are difficulties in accepting such a simple answer to the problem of the
value of science. In any real society there are limited resources
available for the pursuit of all the activities that the State may wish
to engage in. There must be some way of setting up a priority
amongst activities each of which could be admitted to have some
value. For example, it may easily come to the point where, as a
society, we have to decide between building and supporting more
research institutes to pursue the scientific understanding of disease,
and the construction of more hospitals for treating diseases, using
the scientific knowledge that we already possess. The moment we
introduce a priority amongst social goods, the question arises of the
justification for the principle by which that priority was determined.
Such an overall principle must yield some sort of ordering among
the rival activities each of which is a candidate for state support.
Whatever principle we might refer to, it will only serve to assign
priorities if it allows us to rank subordinate goods, for example to
enable us to decide whether advances in medical knowledge are
more to be desired than immediate improvements in health.

In any real society it seems very unlikely that the pursuit of
knowledge would be universally agreed to be the overriding good
to which all other valued activities have to be sacrificed.

The sort of clashes of moral principle that occur can be illustrated by analyzing the objections raised to the unrestricted pursuit
of knowledge in the biological and human sciences. Sometimes we
need to test our biological hypotheses on live animals or to carry
out psychological experiments that entail some kind of suffering on
the part of the people who take part as subjects. Does a gain in
scientific knowledge, however certain it may be, ever take preced-
ence over the welfare of particular sentient beings? This is the sort
of moral issue with which we are familiar from discussions of the

propriety of vivisection. Supporters of the use of live subjects, animal or human, tend to fall back on the utilitarian principle of the welfare of the majority, particularly if the research has medical implications. The opponents of vivisection claim the support of a superordinate moral principle which serves to assess and restrict the application of subordinate moral principles, for example that any advance in medicine is good. A suitable superordinate moral principle would be one which insisted on the intrinsic value and inviolable rights of each sentient being.

A case that has attracted a good deal of attention in the psychological literature is the 'Milgram' experiment. In this study the subjects were deceived into thinking that they were being forced into killing or at least seriously injuring another human being. Many of the subjects suffered greatly as a consequence of what they were forced to do, even though they had volunteered. In such an experiment the director takes it upon himself to deceive those who have agreed in good faith to take part. Milgram defended his conduct by arguing that, though his 'subjects' suffered, nevertheless there was a net gain in that we would now understand how an unjust ruler could maintain control over his subjects. The sufferings of the few bring gains for the many. Opponents of Milgram have argued that it is never right to pursue knowledge that might be beneficial at the cost of sufferings that are certain.

Both the examples illustrate a clash of moral principles. Such clashes are usually resolved by referring to some further consideration that allows us to set one of the clashing principles above the other. This raises the question of whether there might not be some truly basic moral principle to which all such disputes might be referred. I leave it to the reader to try out some candidates, in the confident expectation that, whatever is tried, it will be fairly easy to think up cases where some other principle of seemingly equal moral force can be made to clash with it. We have already seen how the principle that we should pursue the greatest good of the majority clashes with the principle of the intrinsic value of each human being.

It seems that the principle that scientific knowledge is intrinsically valuable will not do as a fundamental ground for justifying scientific research.

But there is another kind of basic or intrinsic good which we might look for in scientific studies. They might be pursued for the beauty, the aesthetic worth of their products. Cosmologists and fundamental physicists have not infrequently written of science as a kind of art. Science is supposed to provide some of the satisfactions that can be derived from practising the more abstract or formal art-forms, such as music. So a society can be expected to support fundamental physics, mathematical cosmology, and the like for just the same kinds of reasons that it can be called upon to support music, literature, and so on. And the justification for such a demand could be looked for in the principle that a society is better in so far as it supports that which is intrinsically good.

The interaction between science and human nature

The kinds of problems that arise when we examine the moral evaluations and judgements that we must make amongst sciences and compare them with other candidates for State support are typical of discussions of the physical sciences. When we come to look at the grounds that might be used to justify public support of the social and psychological sciences we encounter all the arguments concerning derivative and intrinsic worth that I have touched upon already. But a thorough analysis must take account of more subtle considerations. The natural sciences can be pursued independently of the reality they purport to describe. Physicists beliefs about the physical world have no effect on that world. But psychological theories can become part of psychological reality. They can come to be widely believed. As part of the stock of human beliefs they can shape human patterns of thought and feeling. They can filter down to ordinary citizens, who, by believing them, are subtly influenced in how they act. For example, a widespread belief in the Freudian theory of mind will tend to reduce the scope of what people take to be their range of free action. According to that point of view there are forces and influences affecting thought and behaviour that people are usually unaware of and hence cannot control. In the Freudian theory these influences come out of the structures of unconscious thought. Behaviourism, as a popular doctrine, has a similar effect. According to the Skinnerian version of behaviourism the forces that shape human action emanate not from the mind of the actor but from the environment. In radical behaviourism thinking and feeling play no part in shaping action. But there are other theories of the mind which emphasize self-control. A popular contemporary approach is to treat human conduct as the outcome of actors following rules. Rules can be broken and unsatisfactory rules can be changed. If one believes that one's patterns of action are determined by rules rather than by 'mechanistic' processes described in laws of nature one gains greater control over one's life. Belief in such a theory will tend to enhance human powers.

Here we have a much more direct relationship between moral values and scientific research programmes than we have found in the natural sciences. (For a detailed but readily understandable account of this matter see *Images of man in psychological research* by John Shotter). In the case of the physical sciences we can often distinguish between the moral issues that come up in the course of the pursuit of scientific knowledge (vivsection) and those that are involved in the practical applications of that knowledge (germ warfare).

A similar point can be made about sociological studies. If it is true that human societies are in part a product of the social theories and historical beliefs held by their citizens then society is not an independent reality that can be thought of as standing wholly outside a science of society.

But the contrast between natural and human sciences is not so clearcut as these remarks might suggest. Beliefs about the physical world can influence human psychology. It has been suggested that the general acceptance of a sun-centred rather than an earth-centred conception of the universe profoundly changed our sense of the importance of human kind in the scheme of things. Our demotion from the central place was completed by the success of Darwin's evolutionary theory which took away from us our last claim to a special status. As organic beings we were just another product of an essentially natural process.

Science as a social activity

A scientific community is, to some extent, an isolated society with its own social order, its own heirarchies of respect, and its own moral system. This fact has led some sociologists of science, and indeed recently some philosophers of science, to look more closely at the influence that the social structure of the scientific community and its ethical system has on the acceptance and rejection of theories. Recent work has emphasized a parallel between the way a scientific community produces scientific discoveries and the way the larger society produces goods and services.

Economic models

There are various ways in which concepts of economics can be used to understand how a scientific community assesses its products. A broad distinction can be drawn between those sociologists who see science as a market economy and those who see it as more like a medieval guild. These are of course analogies. The main difference between them has to do with how value is arrived at. To understand how value is determined in a market economy we must distinguish between those who produce goods and those who consume them. In our case the goods are the scientific discoveries made by the community of scientists. Effectively in a market economy it is the consumers who set the value of the product. In classical economics value appears in the relation between supply and demand. High value results from a strong demand for a scarce product. This does not seem to capture the way scientists value the discoveries made by their fellows.

A guild economy creates value in a different way. The guild masters, who represent the producers, value the product according to their conception of its worth. This may be related to the skill needed in production, the time that the work takes, or the difficulty of obtaining the proper materials. Ravetz has argued that the scientific community creates values for its products in a manner analogous to the way guilds created value, when they were the dominant social form of manufacturing industry. It is relatively easy to see the scientific community as analogous to a medieval guild. Doctoral students work under a supervisor, more often than not in a research team. The problems they study and the methods they use are very much under the influence of the leader of the project, as in the atelier system in art, when the master planned the overall design and assigned various parts of the picture to his pupils.

But there is another way of making a comparison between an economy and the work of a scientific community. The French sociologist Bourdieu has suggested that scientific institutions and their members accumulate 'reputation'—a kind of symbolic capital. Individual scientists are something like the entrepreneurs of the early forms of industrial society, setting out to build up a reputation by the quality and acceptability of their products, the scientific papers that report their discoveries. Having built up their reputations scientists can use them to create more reputation. The process is something like the accumulation and investment of capital to produce more capital. In this process the type, indeed the quality of the goods turned out is of secondary importance. Latour and Woolgar have explored this analogy in considerable detail (see their *Laboratory life*).

Latour and others have argued that the traditional conception of the scientific community as a group of selfless, dedicated workers producing TRUTH, which is ultimately assessed by reference to the objective criteria of experimental testability, is an inadequate way, indeed a positively misleading way, of describing how a scientific community really works and how it values its products. These analysts have argued that we should understand the scientific enterprise as it is currently carried on without needing to refer to any of the traditional epistemological concepts, such as 'truth', 'falsity', 'knowledge', and so on. Scientific knowledge is not some tested body of truths about how the world works but is the result of a competitive struggle for the ear of the community, waged by the protagonists of various competing points of view by whatever means come to hand, including propaganda, the unscrupulous exercise of power, and skilful use of persuasive rhetoric. According to the sociological reductionists, there is nothing that scientists do that requires anything other than sociological concepts to explain it.

Interesting though this view is, it does seem to ignore certain quite general features of scientific activity. Not least, it overlooks the fact that the scientific community continues to exist and to maintain its place, whatever that is, in relation to the rest of society. If in some way or another the scientific community did not continue to produce the 'goods' in the form of workable, reliable, and usable knowledge, it would hardly have continued to prosper. Furthermore, the achievement of a scientific reputation, good or bad, cannot just be a matter of the exercise of social power in a community. The success or failure of experimental tests cannot be irrelevant to the acceptance or rejection of theories and hypotheses, even if we must admit that no empirical test could determine our attitude to theory.

Paradigms and their changes

A more temperate view of scientific progress has been proposed by T. S. Kuhn in his well-known book *The Structure of Scientific Revolutions*. According to Kuhn we can distinguish two different kinds of phases in the history of a science : he calls these phases 'normal' and 'revolutionary'. Normal science is pursued as long as the metaphysical background and the appropriate methods for conducting scientific enquiries are taken for granted by all the workers in a field. In these circumstances scientists proceed within a 'paradigm', contenting themselves with solving problems which are defined within that paradigm, which also determines what sort of discovery is to count as a solution. During revolutionary phases the paradigms within which scientists have worked comfortably begin to lose their grip on the minds and methods of the community. The metaphysical theory may cease to seem convincing (material atoms may give way to assumptions of fields of force); customary methods may begin to produce anomalous results and provide fewer convincing solutions to problems. The exemplary scientific enquiry which served to embody the paradigm and, as an ideal, lay at the heart of a period of normal science ceases to dominate the practices of the community. A new paradigm appears. Kuhn's most important claim was that the transition from one paradigm to another could not be explained in the rationalistic terms of traditional philosophy of science, for instance by the experimental disproof of a consequence of the older framework. The point of view of those who had worked in the old paradigm was opaque to those who had accepted the new, and vice versa. The paradigm idea is related to the issue we have been discussing above through Kuhn's conception of the scientific community as, roughly speaking, enforcing paradigms. Paradigms are socially promulgated and adherence to the paradigm favoured by a particular group of scientists is one of the conditions for belonging to that community. Kuhn's views have sparked a great deal of discussion, and are by no means universally accepted. (For a detailed treatment of the issues see W. Newton-Smith's *The Rationality of Science*.)

Social influences on the content of science

So far we have been discussing influences on scientific discoveries exerted by moral and political considerations that come from the larger society of which the scientific community forms only a part, and some of the influences which are effective within the community of scientists itself. In both cases we have been talking about the assessment of discoveries, of techniques, and of theories, but not of their content, of what they tell us about man and nature. Beginning with the work of Mannheim, theer has been a growing school of sociologists of science who claim to be able to show not only that there are social influences on the assessment of scientific work, but on content too. For example, the origin of atomism in the seventeenth century, or rather its reintroduction into science, and the appearance of field theories in the nineteenth century should be able to be explained by reference to contemporary social arrangements, class structure, and so on. This idea stems from an insight of Karl Marx's. He realized that many features of the 'superstructures' of social systems—institutions such as the Church, educational programmes, and many other non-economic activities—took their basic forms from the way the productive forces of that society were socially organized in the economy. Marx thought that this influence was concealed from the citizens by an ideology, a theory of their own society shared by all, which systematically provided a false consciousness of how the society was built. It is a short step from attempting to explain the position of the Church in capitalistic societies along Marxist lines to trying to explain the rise of certain scientific ideas by reference to the facts of economic organization. The most important distinguishing feature of the social organization of production, according to Marx, was class position. It was through their class position that people's ideas were influenced by the economic order.

As it stands this is too general a thesis to be convincing as an explanation for the origin and spread of specific scientific ideas. Sociologists of science have tried to work out in more detail just how a person's class position might play a role in influencing the kind of theories that sprang to his mind, so to speak. However, it has proved extremely difficult to establish a direct causal link between the productive arrangements of a society and the content of its science. David Bloor, in *Knowledge and Social Imagery*, has suggested a rather more complex set of relationships than are to be found in classical Marxism. In his view there is a shared framework of thought, which he calls a general ideology, characteristic of a society at some historical moment. The general ideology expresses itself both in preferences among theoretical ideas on the part of scientists, and in the class interests of the citizens, which are revealed in how they build up the economic organization and superstructural institutions of that community. Scientific preferences will lead to the appearance of certain sorts of ideas which, looked at rather broadly, will be seen to be correlated with the class interests of the scientists involved. This theory has the advantage of not requiring anything so mysterious as a direct positive causal relationship between the class interests of scientists and the ideas they favour. Both are seen as realizations of something deeper.

As philosophers of science we need to ask ourselves whether studies of the kinds I have been illustrating in this chapter are so complete as accounts of what scientists think and do that the traditional philosophical concerns expressed in the study of logic, metaphysics, and epistemology should be of no further interest. Indeed, perhaps the very philosophies of science that we have looked over in this book are themselves nothing but refracted images of the class interests of those who favoured them. To answer this sort of question is beyond the scope of this study, but it is not difficult to see that if the 'strong programme' of sociological reduction of all epistemological issues were to succeed it would have to overcome a very deep objection. This is the accusation of having committed the 'genetic fallacy'. It does not follow that because one has given a correct account of how some belief came to be held that we are not entitled to ask about its truth as well. Someone may come to believe something that is true because he is frightened of a teacher, influenced by the attractive personality of a friend, or because he is a creature of his class position. The revelation of how that belief was caused has no bearing on its value as knowledge.

Summary of the argument

1. *Science in society*
   (*a*) Science as an activity justified by the value of its applications
   According to this view scientific research is neutral and moral issues arise only in its applications.
   (*b*) Science as an activity justified by its intrinsic worth
      (i) Scientific knowledge is good in itself. But it is easy to devise cases where the claim clashes with equally attractive alternatives.

    (ii) Unrestricted pursuit of scientific knowledge can be evil.

    (iii) The defence of the value of science as an art form avoids some of these difficulties.

(c) The interaction between science and human nature

    (i) Psychological and sociological beliefs are part of the causes of human behaviour.

    (ii) Theories in the physical sciences affect our view of ourselves indirectly.

2. *Science as a social activity*

(a) Economic models

    (i) A market economy is a poor analogy for science as productive work.

    (ii) The medieval guild, which valued its own products, is more illuminating.

    (iii) According to another model reputation can be treated as capital.

(b) Paradigms and their changes

A cluster of metaphysical assumptions, empirical methods, and exemplars of good work serves as a paradigm. Normal science is research within a paradigm but in revolutionary periods paradigms are changed, but not in rational ways.

(c) Social influences on the content of science

    (i) According to the Marxist thesis the content of scientific theories, as part of the superstructure of society, could be a product of the class position of scientists and so economically determined.

    (ii) A more subtle thesis has been proposed by Bloor; that a general ideology underlies both scientific thought and the forms of material production.

Further reading

THE ITEMS ARE arranged under chapters, and are chosen to provide reading which supplements and extends the material therein.

CHAPTER 1

S. Toulmin, *The Philosophy of Science* (London: Hutchinson, 1967).

M. Polanyi, *The Tacit Dimension* (New York: Doubleday, 1966).

S. Toulmin, *Foresight and Understanding* (London: Hutchinson, 1961).

J. T. Davies, *The Scientific Approach* (London and New York: Academic Press, 1965).

R. Harré, *An Introduction to the Logic of the Sciences* (London: Macmillan, 1965; Papermac 137).

P. B. Medawar, *Pluto's Republic* (Oxford: University Press 1982).

CHAPTER 2

A. Koestler, *The Sleepwalkers* (London: Hutchinson, 1959).

J. S. Mill, *A System of Logic* (London: Longmans Green, 1879).

P. Edwards (ed.), *The Encyclopedia of Philosophy* (New York: The Macmillan Co., Free Press, 1967), Vol. 6 pp. 289–96.

J. O. Wisdom, *Foundations of Inference in Natural Science* (London: Methuen, 1952).

R. Von Mises, *Positivism* (Cambridge, Mass.: Harvard University Press, 1951).

R. Carnap, *Philosophical Foundations of Physics* (New York: Basic Books, 1966).

K. R. Popper, *The Logic of Scientific Discovery* (London: Hutchinson, 1962).

K. R. Popper, *Conjectures and Refutations*, (London: Routledge and Kegan Paul, 1963).

A. J. Ayer (ed.), *Logical Positivism* (Glencoe, Ill.: Free Press, 1959).

Chapter 3

O. Neugebauer, *The Exact Sciences in Antiquity* (New York: Harper, 1962).

A. Pannekoek, *A History of Astronomy* (London: George Allen and Unwin, 1961).

D. M. Knight, *Atoms and Elements* (London: Hutchinson, 1967).

W. H. Brook (ed.), *The Atomic Debates* (Leicester: Leicester University Press, 1967).

A. S. Eddington, *New Pathways in Science* (Cambridge: Cambridge University Press, 1938).

G. Berkeley, *A Treatise concerning the Principles of Human Knowledge* (Everyman's Library No. 483, London: Dent, 1946).

P. W. Bridgman, *The Logic of Modern Physics* (New York: Macmillan, 1954).

A. C. Benjamin, *Operationism* (Springfield, Mass.: Thomas, 1955).

R. M. Blake, C. J. Ducasse, and E. H. Madden, *Theories of Scientific Method* (Seattle Wash.: University of Washington Press, 1960).

E. Mach, *The Analysis of Sensations* (New York: Dover, 1959).

E. Mach, *The Science of Mechanics* (trans. T. J. McCormack) (La Salle, Ill.: Open Court, 1960).

N. R. Campbell, *The Foundations of Science* (New York: Dover, 1957).

J. J. C. Smart, *Philosophy and Scientific Realism* (London: Routledge and Kegan Paul, 1963).

Chapter 4

E. A. Burtt, *The Metaphysical Foundations of Modern Science* (London: Routledge and Kegan Paul, 1964).

E. E. Harris, *The Foundations of Metaphysics in Science* (London: George Allen and Unwin, 1965).

M. Capek, *The Philosophical Impact of Contemporary Physics* (New York: Van Nostrand, 1961).

R. Swinburne, *Space and Time* (London: Macmillan, 1968).

D. Hume, *An Enquiry Concerning Human Understanding* (ed. L. A. Selby-Bigge), 2nd edn. (Oxford: Clarendon Press, 1902).

M. Bunge, *Causality* (Cambridge, Mass.: Harvard University Press, 1959).

W. D. Ross, *Aristotle* (London: Methuen, 1949).

W. A. Wallace, *The Scientific Methodology of Theodoric of Freiberg* (Fribourg, Switzerland: The University Press, 1959).

R. J. Boscovich, *A Theory of Natural Philosophy* (Cambridge, Mass.: M. I. T. Press, 1966).

I. Kant, *The Metaphysical Foundations of Natural Science* (translated J. Ellington)(Indianapolis, Ind., and New York: Bobbs-Merrill, 1970).

L. Pearce Williams, *Michael Faraday* (London: Chapman and Hall, 1965).

CHAPTER 5

G. Schlesinger, *Method in the Physical Sciences* (London: Routledge and Kegan Paul, 1963).

B. Schonland, *The Atomists (1805–1933)* (Oxford: Clarendon Press, 1968).

R. E. Peierls, *The Laws of Nature* (London: George Allen and Unwin, 1955).

B. Ellis, *Basic Concepts of Measurement* (Cambridge: Cambridge University Press, 1968).

CHAPTER 6

C. G. Hempel, *Aspects of Scientific Explanation* (New York: The Free Press, 1965).

H. Herz, *The Principles of Mechanics* (New York: Dover, 1965), pp. 1–41.

M. B. Hesse, *Models and Analogies in Science* (London: Sheed and Ward, 1963).

P. Duhem, *The Aim and Structure of Physical Theory* (trans. P. Wiener), (Princeton, N.J.: Princeton University Press, 1954).

I. Scheffler, *The Anatomy of Inquiry* (New York: Knopf, 1963).

R. Harré, *The Principles of Scientific Thinking* (London: Macmillan, and Chicago, Ill.: University of Chicago Press, 1970.

CHAPTER 7

D. Bloor, *Knowledge and Social Imagery* (London: Routledge and Kegan Paul, 1976).

R. Harré (ed.), *Problems of Scientific Revolution* (Oxford: Clarendon Press, 1975).

T. Kuhn, *The Structure of Scientific Revolutions* (Chicago: Chicago University Press, 1962).

B. Latour and S. Woolgar, *Laboratory Life* (Los Angeles: Sage, 1979).

W. Newton-Smith, *The Rationality of Science* (London: Routledge and Kegan Paul, 1981).

J. R. Ravetz, *Scientific Knowledge and its Social Problems* (Oxford: Clarendon Press, 1971).

J. Shotter, *Images of Man in Psychological Research* (London: Methuen, 1975).

Index

OXFORD

MORE OXFORD PAPERBACKS

Details of a selection of other books follow. A complete list of Oxford Paperbacks, including The World's Classics, Twentieth-Century Classics, OPUS, Past Masters, Oxford Authors, Oxford Shakespeare, and Oxford Paperback Reference, is available in the UK from the General Publicity Department, Oxford University Press, Walton Street, Oxford, OX2 6DP.

In the USA, complete lists are available from the Paperbacks Marketing Manager, Oxford University Press, 200 Madison Avenue, New York, NY 10016.

CHARLES DARWIN AND T. H. HUXLEY: AUTOBIOGRAPHIES

Edited with an Introduction by Gavin de Beer

Charles Darwin and his 'Bulldog', T. H. Huxley, are presented here as each depicted himself. Two men of completely different temperament, they had immense admiration and respect for each other.

'It is singularly appropriate that these two autobiographies should be reprinted together, for in the history of science there can hardly have been a more fruitful and essential association of two men of such strikingly different personalities.' *Journal of Natural History*

'the fragmented autobiography which Darwin wrote for private circulation among his family is one of the most charming documents I have read in years' Benny Green, *Spectator*

DARWIN

Jonathan Howard

Darwin's theory that men's ancestors were apes caused a furore in the scientific world and outside it when *The Origin of Species* was published in 1859. Arguments still rage about the implications of his evolutionary theory, and scepticism about the value of Darwin's contribution to knowledge is widespread. In this analysis of Darwin's major insights and arguments, Jonathan Howard reasserts the importance of Darwin's work for the development of modern biology.

'Jonathan Howard has produced an intellectual *tour de force*, a classic in the genre of popular scientific exposition which will still be read in fifty years' time.' *Times Literary Supplement*

Past Master

GENESIS
The Origins of Man and the Universe
John Gribbin

The author begins his cosmic history some fifteen billion years ago, a split second after the 'big bang' of creation, and leads us on a fascinating voyage through vast reaches of time and space to the here-and-now of life on earth today.

'A splendid book.' Douglas Adams, author of *The Hitch Hiker's Guide to the Galaxy*

'Britain's answer to Carl Sagan? Comparisons between this latest work by John Gribbin and the much publicized *Cosmos* are inevitable and the comparison I found favourable to *Genesis*.' *Physics Bulletin*

GAIA
A New Look at Life on Earth
J. E. Lovelock

De Lovelock's Gaia hypothesis first took the scientific world by storm in the mid-seventies. He proposed that all living things on the earth are part of a giant organism, involving air, oceans, and land surface, which for millions of years has controlled the conditions needed for a healthy planet. While stressing the need for continued vigilance, Dr Lovelock argues that, thanks to Gaia, our fears of pollution-extermination may be unfounded.

'This is the most fascinating book that I have read for a long time.' Kenneth Mellanby, *New Scientist*

THE EXPANDING CIRCLE
Ethics and Sociobiology
Peter Singer

Where do ethical standards come from? Are our notions of good and evil created by reason, or by evolution? Can society shape its own destiny, or must it merely reflect biological imperatives? In answering these questions Peter Singer (author of the widely acclaimed *Animal Liberation* and, with Deane Wells, *The Reproduction Revolution*) is particularly concerned with the light thrown on our morality by the new science of sociobiology. He builds up a convincing picture of an ethical system which, though biologically grounded, has expanded from this base to become more rational and objective.

'Unwaveringly clear, rigorously accessible.' *Sunday Times*

AN INTRODUCTION TO THE STUDY OF MAN

J. Z. Young

There are many ways of approaching the study of Man. Professor Young believes that biological knowledge provides a useful framework to help us to understand ourselves. Modern biology embraces many disciplines, and in this book a synthesis is made tracing the sources of human activity from their biochemical basis to the highest levels of consciousness.

'Professor Young sticks to straight and informative science . . . is rivetingly interesting, and conveys a constant sense of the controlled, critical curiosity which is what science is about.' *Guardian*

'an impressive performance' *Observer*

A SHORT HISTORY OF SCIENTIFIC IDEAS
TO 1900

Charles Singer

This book places the basic scientific ideas developed by man in a framework of world history, from the earliest times in Mesopatamia and Egypt until A.D. 1900, and treats not only the physical and chemical but also the biological disciplines. Published over 20 years ago to glowing reviews, it has become a standard work.

'One reason why this new history of science is assured of an illustrious career is that it is a work of such consummate art . . . masterly in conception and execution.' *New Scientist*

'this book is in the very front rank' *Advancement of Science*

'Dr Singer deserves well of Western man' *The Economist*